FIRING

THE STEAM LOCOMOTIVE

PREFACE

The importance of the work performed by the firemen of the steam locomotives is recognized and appreciated, not only by the firemen's supervisors and officials of the railroad, but by business men and the general public as well. It is acknowledged that the progress of our American Railroads contributed largely towards keeping the arteries of our transportation system open, promoting our industrial development, raising our general standard of living, to mention only a few of the factors.

We cannot afford to sit back and rest on our laurels. We must advance or we will find ourselves moving backwards, losing ground to opposing forces. It is of utmost importance that the men who choose to enter upon a railroad career and start in by firing locomotives should be thoroughly trained and familiar with their responsibilities, duties and the equipment they will have to handle.

This manual is prepared and offered to the employes of the Reading Company for the purpose of assisting them in fulfilling their important responsibility by presenting them with the best established method of firing practice together with descriptions and functions of various items of equipment they must handle. The material in this manual should prove equally useful to the experienced firemen as well as the men who more recently have undertaken the responsibility of firing the locomotive. It offers an opportunity to all to become acquainted with the requirements of their job and adopt a uniform practice based upon operating experience which represent some of the best established practices in locomotive firing.

ACKNOWLEDGMENT

In the preparation of this manual, in order that it may be as nearly complete as possible, the assistance and advice of numerous consultants has been solicited and the fullest use made of existing materials applicable to the subject.

The contents were compiled and arranged by R. A. Reeder, Superintendent Fuel and Locomotive Performance, Reading Company, with the cooperation of various officers and staff members of the Reading Company.

A complete listing of all companies and individuals whose ideas, suggestions, and technical information have been included, or who have otherwise assisted would not be feasible. The value of their contributions is recognized and appreciation expressed for their splendid assistance and cooperation. To all who have in any way aided, grateful acknowledgment of their contributions is made and to them this manual is presented with the hope that it will be of service and will assist in increasing the efficiency of those engaged in the important work of firing steam locomotives.

Considerable information contained in this manual is based on instructions prepared by manufacturers of the various locomotive equipment, "The Traveling Engineers Examination Handbook", and "Good Firing Practice Manual" prepared by the New York, New Haven and Hartford Railroad.

ENDORSEMENT

READING COMPANY

Philadelphia, Pennsylvania

November 16, 1947

To Engine Service Personnel:

As a part of our employes training program, this manual has been prepared to assist you in your various duties as firemen and enginemen.

We have endeavored to provide, in handy reference form, the complete facts on good steam locomotive firing.

_____
Supt. of Motive Power

_____
General Manager

_____
Chief of Personnel

_____
Vice President

—————————————————— **ENDORSEMENT**

November 16, 1947

Mr. N N Baily,
Vice President, Operation and Maintenance,
Reading Company,
12th & Market Streets,
Philadelphia 1, Pennsylvania

Dear Mr. Baily:

Having read this manual on "Firing the Steam

Locomotive", we endorse it as a very informative and

practical manuscript.

Our enginemen and firemen should find this book a

valuable asset in their routine of duty.

Sincerely,

Wm. R. Hamm   L. C. Lobach
—————————————   —————————————
General Chairman   General Chairman
Brotherhood of Locomotive   Brotherhood of Locomotive
Engineers   Firemen and Enginemen

TABLE OF CONTENTS

SAFETY

SAFETY

"Safety is of the first importance in the discharge of duty" and "Obedience to the rules is essential to safety."

The most important factor in the success of railroad men, regardless of their job, is SAFE WORK HABITS. Not only knowing the safe way of doing your work, but practicing safety until it becomes second nature just as lacing your shoes or walking along the street is done, with the least exertion of mental effort. There may be occasions when you will not have time to think about what you want to do. You must learn the right, safe way of doing your work until you can do it instinctively—without thinking.

It is not the intention of this manual to go into detail and list all the individual items and safety factors the locomotive fireman must cultivate and practice. There is a manual issued for that specific purpose and every fireman must be thoroughly acquainted with it. However, it is intended to bring out some of the more serious factors confronting the locomotive fireman and very definitely establish the necessity to work safely and practice safety.

The emphasis on safety should not be confined to the safeguarding of equipment and engineering revisions but should be placed on the real heart of the problem—the human factor. Here is the job, carrying out essential duties, and in doing his work the fireman must recognize there are dangers which he must continually face. They are part of the job. However, unnecessary chances play no part in the efficient performance of his job. They mar it or delay it; they may be the cause of forcing him to lay it down forever. There is a difference between the risk a person takes to save another's life or to overcome great dangers in order to gain recognition for some high or noble purpose and the foolish risks which mean throwing away one's life or one's future usefulness, for nothing.

Let's put the problem squarely in your lap. Look over and carefully study your Safety Manual. Critically judge the importance of every article and use such principles as you can honestly accept to the end that greater safety for body and property may be insured.

The whole question on safety will be largely determined by the reactions of individuals to new situations. In order to be safe we must be observant, alert, and quick to think; we must possess or develop proper coordination of the brain and muscle, a genuine sense of proportion, good judgment and the ability to control ourselves under great strain or emotion. There is a right way of doing every job and the right way is the safe way. It therefore becomes more important and apparent that the fireman learns to do his work correctly.

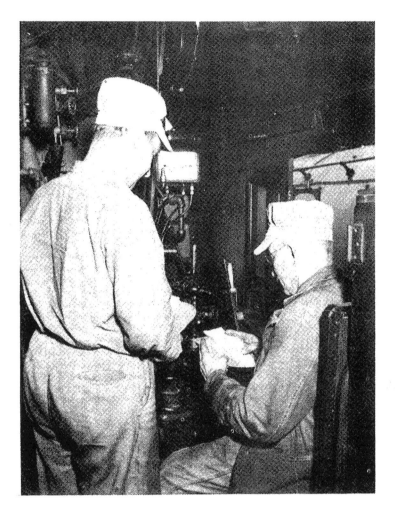

Read your orders carefully.

Correct way to get on and off an engine.

Keep the deck clean.

Inspection and service essential for safe and economical operation.

SAFETY (Continued)

The most important physical faculty a railroad man possesses is good eyesight; therefore, every precaution should be taken to guard against eye injury.

Records seem to indicate that eye injuries occur more frequently to railroad men than any other form of accident. Some precautions you must take to avoid these accidents are:

1. If a passing train on an adjoining track is approaching, it is good policy to turn your back towards the train to avoid cinders from entering your eye.

2. Slack or dry coal is also likely to fly around and will probably blow into your eye; therefore, wet the coal down sufficiently to keep it from flying around.

Another form of accident that has had serious results to locomotive firemen is getting caught in the stoker conveyor screw. This is a mechanical device with sufficient power to crush even hard lumps of coal and such a part of your body as a foot or a hand will not stop the conveyor screw. Therefore, before attempting to remove any obstructions from conveyor unit be certain to: (1) that the stoker steam valve and booster valve are shut off, (2) that the stoker engine is in the neutral position before attempting to remove any obstructions from the conveyor unit, (3) that the slides are over the conveyor trough before stepping over the conveyor screw.

Develop the habit of facing the engine when getting off. This permits the use of both hands on the grab iron or hand hold until your foot, or feet have reached a firm footing and insures the only safe manner of getting off. Other methods of getting off the engine are not only awkward and inconvenient but are dangerous and invite accidents.

Safety chains are provided on both sides of the locomotive, between the cab and the tank. When the locomotive is in motion, this chain should be hooked in position to protect the engine crew from being thrown off the deck of the locomotive, if it should make a sudden lurch. When not in use, the chain should not be allowed to dangle but should be hooked up on the cab.

Good housekeeping is always an indication of safe work habits. Keep your deck clean so that you are free to move around without danger of tripping over large lumps of coal, your firing tools, or other objects; and don't be afraid to use the sprinkler hose.

Loss of fuel and personal injuries can be avoided by keeping the coal properly trimmed on locomotive tenders.

REMEMBER

The Operating Rules, theoretically, is the air railroad men breathe. Their purpose is to make train service safe as well as making it safe for the employes. Conforming to these rules is the reason you are on the job. A hundred years have gone by in the preparation of the Operating Rules since the railroads have been in existence. Don't get the impression these are the only safety rules. Be sure you understand your Safety Rules and your Operating Rules thoroughly.

Be sure to LOOK where you are going.

CREW COOPERATION

Reading Lines

CREW COOPERATION

INTRODUCTION

The word COOPERATION is too big and meaningful to be completely defined. Every crew recognizes the importance of COOPERATION, but, nevertheless, every once in a while someone violates that principle. Lack of cooperation makes the job tougher for everyone concerned. The usual result is, as everyone knows, trouble, and sometimes even failure. This is true in every line of work, whenever a group is involved.

Perhaps there is no other situation where COOPERATION between two men is more important than it is between the crew in the cab of a locomotive. Upon the performance and decision of these two men depends the safety of the passengers and crew and the care of valuable freight and railroad equipment.

The engineer is responsible for the locomotive and its performance on the road. The fireman is responsible to the engineer. A recognition of this relationship and a willingness to COOPERATE will promote teamwork which will get the best possible results under any condition.

The average person does not realize how much experience, information, and teamwork are necessary to bring about an "on time" performance.

One steam failure, regardless of cause, may tie up a whole section of the railroad system. Sometimes the cause may be traced back to some other condition, but there can be no excuse for lack of crew COOPERATION or poor firing practice.

There are many ways in which ALL individuals responsible for the operation of a locomotive can help each other. Good COOPERATION will help each individual to do his job better and will get the best performance and efficiency out of each locomotive with the least effort.

Cooperation In Firing the Locomotive

HOSTLER AND ROUNDHOUSE CREW

At this point the results of COOPERATION begin to have very important effects. The fire builder, the engine watchman, and the hostler should make sure that the locomotive is turned over to the engine crew ON TIME with a good, well-coked fire, not over two gages of water, and steam pressure within fifty pounds of the maximum boiler capacity. Give the engine crew a good start.

The inspection and service a locomotive must receive after each run is essential for safe and economic operation of the equipment. However, unnecessary delays at the round house may keep the locomotive off scheduled runs and force the engine dispatcher to substitute a locomotive not exactly fitted for the job that has to be done. The usual result is a dissatisfied engine crew at the start of the run, additional work for the fireman during the run, and finally winding up in unnecessary delays which not only bring the train in late but may tie up a whole section of the railroad system.

THE ENGINE CREW

It is, of course, understood that the engineer has charge of and is responsible for the work of the fireman while he is under his charge. Consequently, in the interest of fuel economy and smooth operation of their locomotive, it is very important for him to keep the fireman fully informed on any changes that may affect the steam pressure requirements.

Primarily the engineer is interested in maximum horsepower, that is, horsepower at speeds; however, a considerable amount of steam can be saved by shortening the cut-off whenever possible. On the other hand, if the cut-off is to be lengthened or the throttle is widened, the fireman should be informed of these changes.

The engineer should avoid slipping, if possible, as this has a tendency to tear holes in the fire.

The fireman should COOPERATE with the engineer by giving him a report of any defects that have been discovered so that they can be included in the report. A complete, intelligent report will assist the repair department to maintain the locomotive in good condition.

The fireman should notify the engineer of any bad conditions in the firebox so that he can work the engine accordingly until the trouble can be corrected.

The fireman should anticipate the need for more or less steam and control his fire accordingly. When the train is approaching a long pull, a long down grade, stops, or other conditions that will affect the steam requirements, the amount of coal delivered to the firebox should be regulated to meet the requirements.

FIREMAN'S DUTIES

FIREMAN'S DUTIES

INTRODUCTION

An efficient fireman is a man who possesses the skill and knowledge which enable him to burn no more fuel than necessary and to make the fuel which he supplies to the firebox burn so hot that it evaporates into steam as much water as possible. Briefly, he makes the fuel perform its maximum duty.

Of course, there are other essential characteristics which increase the value of a man as a fireman, but, the ability to keep up steam is the first consideration. A systematic division of the fireman's duties will assist him in checking the important factors which will help him to be a successful fireman and do his work easily and satisfactorily.

The following list way appear to be long and impressive. However, each item is an essential part of the fireman's job which he usually does almost instinctively. Follow these lists and practice each item until it becomes a habit. CORRECT PRACTICE DEVELOPS GOOD WORK HABITS. The reason some people are never too successful in what they are doing, regardless of how long they work at it, it because they did not get started doing their work correctly. Then they kept on working and practicing their mistakes.

FIREMAN'S DUTIES ON ARRIVAL AT ENGINE-HOUSE

1. Report on time to enginehouse dispatcher or foreman.
2. Consult the bulletin board.
3. Compare watch with standard clock.
4. Register on prescribed form.

FIREMAN'S DUTIES AFTER ASSIGNMENT TO LO-COMOTIVE

1. Check the amount of water.
2. Check the steam pressure.
3. Examine the fire.
4. Check the condition of the brick arch.
5. Build up the fire with a SCOOP before starting out so that it will be in good condition.
6. Blow out water glass and gage cocks.
7. Compare gage cocks with water glasses.
8. See that water, flagging equipment, and all necessary tools and equipment are provided.

9. Check and observe the grade, size, and condition of the coal supply.
10. Examine the ash-pan to see that it has been properly cleaned out and no green coal is left there which might ignite and warp the sheets. See that all ash-pan openings are closed and secured.
11. Inspect the stoker.
12. Perform any other duties assigned by the engineer.

FIREMAN'S DUTIES ON THE TRIP

1. Observe all safety precautions.
2. Start stoker operation and maintain a good fire.
3. Check fire frequently.
4. Check size and condition of coal frequently.
5. Watch, and call, all signals.
6. Maintain proper steam pressure.
7. Plan ahead, and fire according to the need for steam.
8. Avoid black smoke.
9. Keep deck clean.

FIREMAN'S DUTIES AT END OF TRIP

1. Ease off on firing before arrival at terminal so that fuel will not be wasted, but still maintain sufficient fire to keep the steam pressure up until the hostler takes charge. If possible, get fire in shape by hand firing.
2. Before arriving at the terminal at the end of the trip, shut off the stoker by putting the stoker-operating lever in neutral position and by closing the stoker-engine throttle valve and jet valves. The leakage ports will furnish sufficient steam to protect the steam jets and the distributing plate.
3. Report to the engineer any defects in the boiler, fire-box, stoker, water supply system, or any other defects that affect the performance of the locomotive or may cause a mechanical failure.

CAB CURTAINS

Cab curtains are installed on each side and the back of all steam locomotives as a protection for the engine crew against bad weather. When not in use, they should be pushed to one side and FASTENED in the open position.

The fireman should exercise reasonable care to avoid any damage to the cab curtains which would render them ineffective when the need for their use should arise

COAL

Reading Lines

Mining Bituminous Coal in West Virginia

23

Men pick out the impurities as the coal moves over the shaking conveyor in the tipple. Mechanically screened for size, lumps move forward to railroad cars waiting below. Fuel in the lower conveyor already is sliding into its chute. Slack coal, first to drop, does not show in this photograph.

In an Appalachian field, cars receive loads according to various sizes. Through the mountainside head house, a belt conveys coal from the mine.

COAL

Coal has been a subject of many intensive studies not only by the United States Government Bureau of Mines but industrial organizations and educational institutions as well. It cannot be hoped to cover the subject here thoroughly but it is the propose here to give a better understanding of some of its principal characteristics.

CLASSIFICATIONS

Coal is a product of ancient vegetation which decomposed and, under great pressure, formed the various kinds of coal. There are several methods of classifying or rating coal. One classification is based on the ratio of carbon to volatile matter, another on the method or rate of burning, and another on the size of the pieces sold for use. Based on the type or nature of the coal, the United States Geological Survey classifies native coal into the following groups: peat, lignite, sub-bituminous cannel, bituminous, semi-bituminous, semi-anthracite and anthracite.

Coals are also classified according to their tendencies to react to heat and their tendency to disintegrate and decompose. The so-called CAKING COALS show a tendency to melt and run together when heated. These coals swell when heated and may give rise to either soft or hard clinkers. COKING COAI swells when heated but does not cake as freely as caking coal. FREE BURNING COAL swells little when heated and it does not cake or coke. It burns freely and sometimes even flashily. Other classifications apply more specifically to the so-called hard coals.

BITUMINOUS COAL

Bituminous coal is commonly called soft coal. When it is high in volatile matter it burns rapidly. It usually comes as "the run of the mine", in lump and slack form. It is mined in about twenty different states, but most of it comes from the East. Unless fired properly, it will give off large quantities of smoke which also carries much of the heat value of the coal with it.

CARBON AND VOLATILE MATTER

Coal is composed of two main heat-producing elements: carbon and volatile matter. Carbon is contained in the solids, the part of the coal which remains on the grates. It contains the greatest portion of the heat producing matter. Volatile matter is composed of the gases released from the coal when it is heated to a certain temperature. Some of these gases thus released are hydrogen, methane and carbon monoxide.

At a proper temperature these gases will burn and they constitute an important element for producing heat. The finest Eastern bituminous coal may rate as high as forty per cent in volatile matter.

INSPECTING THE COAL IN THE TENDER

Knowing that there is such a large difference in coal, when it comes from different mines of the same seam, it is well for the fireman to know what causes certain conditions in the firebox. This knowledge, together with the skillful application of the rules of good firing will enable the fireman to get the best possible results obtainable from the coal he has in his tender.

The fireman cannot tell by looking at the coal in the tender how well it will burn in the firebox. He must burn the coal to find out how well it will burn. There are certain conditions that can be observed which will help to get the greatest efficiency out of coal. Some of these conditions can be observed in the tender; others can be observed in the firebox.

WHAT TO LOOK FOR IN THE TENDER

Before starting out on a run the fireman should make the following observations of the coal in his tender:

1. Observe the coal in the tender to see whether it is lump or slack.
2. Observe the condition of the coal to see whether it is wet or dry.
3. See that the coal is properly trimmed to avoid the possibility of coal falling off and injuring someone and to avoid wasting coal.

The fireman should also try to keep the coal supply free from foreign matter and remove any substance which might interfere with proper burning or get jammed in the stoker.

While on the trip, the fireman should continue to observe the condition of the coal occasionalsy so as to determine whether it is running wet or irregular in size. These two factors are important for most efficient burning of coal.

WHAT TO LOOK FOR IN THE FIREBOX

As it has been stated previously, coal varies widely in many respects. Some grades of coal tend to clinker badly; others form a large amount of ash, and still others might produce an excessive amount of smoke. If the fireman knows that different grades of coal do have these tendencies, he can look for and learn to recognize the different conditions when he inspects the fire. He can then take measures, in controlling his fire, to avoid trouble or at least to offset the difficulty well enough to get acceptable results.

COMBUSTION

The brick arch is necessary because a large quantity of gases are given off during the process of combustion; the arch helps to burn these gases more completely by forcing them back over the bed of the fire. This action thoroughly mixes the air and the gases before they enter the flues and this promotes good combustion, saves fuel and prevents black smoke.

Perfect condition of Fire Bed, level throughout the Middle
—but thicker along the sides.

Wrong condition of fire bed. Too deep in the middle where
the draft is weakest—and too shallow at the sides where the
draft is strongest.

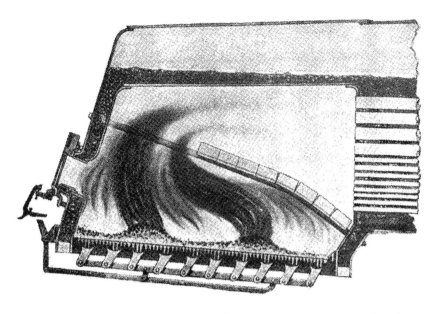

Two BANKS in the fire. Either the fireman carelessly put too much coal on the parts of the fire where the banks form — or CLINKERS prevent access of sufficient a r through those parts to burn the coal placed there.

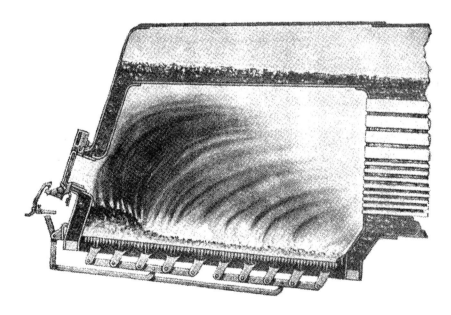

Imperfect condition of the fire. The Bank—"Heel"—blocks off that part of the fire and causes poor combustion resulting in engine emitting black smoke.

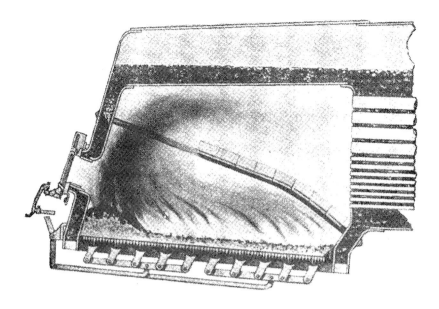

Imperfect condition of the fire. The BANK just beneath the fire door has put that part of the fire out of action. It is an enemy to the fire, the fireman and the engine.

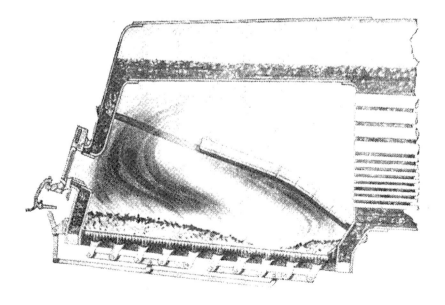

Imperfect condition of the fire. The hole near the front was probably caused by shaking the grates too hard under this part of the fire bed.

Perfect condition of fire bed — LENGTHWISE. Level and uniform AII OVER.

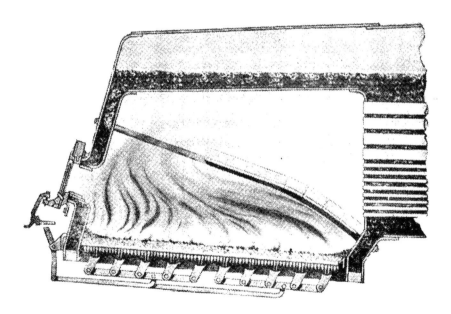

How an ARCH compels the gases escaping from the coal to MIX together and burn before they can escape from the firebox.

COMBUSTION

Technically defined, combustion is the rapid oxidization caused by a chemical union of a substance with oxygen and accompanied by the production of light and heat. More plainly stated, as it applies to locomotive firing, it means the act or state of burning to produce heat.

To produce heat in a locomotive firebox THREE CONDITIONS are necessary—and only three. First, there must be a supply of fuel. Second, there must be an adequate supply of AIR. Third, the fuel and the air must be brought together at a temperature at which they will burn — the IGNITING TEMPERATURE.

COAL

The composition of coal has already been discussed. It will, however, bear repeating that coal is composed of coke and volatile matter (gases). When coal is thrown on the fire, the gases are driven off by the heat and the coke remains on the grates. Both coke and the gases will burn, but before any burning can take place they must be supplied with air at what is known as the igniting temperature.

AIR

Air is composed of several gases but it is generally recognized as a mixture (by volume) of approximately eighty percent nitrogen and twenty per cent oxygen. Nitrogen is an inert gas and takes no part in combustion. It merely dilutes the oxygen and the products of combustion and absorbs heat, thereby making the intensely active oxygen safe to use and keeping the firebox temperatures within practical limits.

IGNITING TEMPERATURE

The igniting temperature is that at which a fuel may begin to burn, that is, combine with the air (oxygen) and generate heat. Of course, this temperature is different for different kinds of fuel.

The igniting temperature required to burn some of the gases in a locomotive firebox is about 1800°. On the other hand, the highest temperature that can be obtained is about 2800°. Therefore, there is only a comparatively small range between the maximum temperature that can be obtained and the point at which the gases would no longer burn because the temperature of the firebox would be below the igniting temperature of the gases.

It is important to keep this fact in mind. Burning will not take place unless the igniting temperature is present. This explains why it is necessary for the locomotive fireman to carry a bright fire as one of the requirements for properly firing a locomotive.

RATE OF COMBUSTION

The element of time, or the rate of combustion, is an important factor. The rate of combustion in a locomotive firebox per square foot of grate area per unit of time is relatively high. Coal must be burned under a forced draft to evaporate enough water into steam by the heat liberated from the fuel. If the fire is not hot enough to burn the fuel rapidly, part of it remains unburned and passes out of the stack as black smoke, and is wasted.

The temperature in a properly drafted and properly fired coal-burning locomotive should range between 2000° and 2500° at the center of the firebox. The temperature at the side and end sheets is lower because the temperature of the water on the other side of the sheets is lower than the temperature in the firebox.

It is desirable to develop as high a temperature as possible along the side and end sheets in order to obtain the highest efficiency in steam generation.

FACTORS AFFECTING GOOD COMBUSTION

Since it is good combustion that gets the heat out of coal to produce steam, some factors bearing on this subject are presented here.

The fire in the firebox should be as light on the grates as the work done by the engine will permit. The fire should be evenly distributed over the entire grate area and should be free from banks and clinkers.

When coal is put into the firebox, the heat from the fire drives the volatile combustion gases from the coal. These gases are ignited as they pass through and over the BRIGHT fire, leaving coke to burn on the grates. To burn the gases, a BRIGHT, WHITE almost incandescent coke fire is necessary. Under good conditions the gases will be burned a short distance before they enter the flues. If the gases are not consumed in the firebox, they will burn only a short time after they enter the flues of the boiler, because the water surrounding the flues, absorbs so much heat that the temperature of gases is lowered below the igniting point. Thus, most of the gases together with the fine particles of carbon carried along by the draft, that are not burned before they leave the firebox, are wasted.

BLACK SMOKE, at the stack consists of this unburned fuel, and indicates incomplete combustion. Production of black smoke can be controlled by careful and intelligent firing.

Fresh coal should be added to the fire in comparatively small quantities. In the first place, opening the fire door to add fresh coal allows a considerable amount of cold air to enter the firebox, thus reducing the firebox temperature. In the second place, this fresh coal has to be "torn down" into its elementary parts BEFORE BURNING can commence (which also absorbs a considerable amount of heat). Therefore, a large amount of coal on the fire at one time will deaden the fire and lower the temperature of the firebox considerably. It will have the same effect as putting a blanket over the fire.

Gases are released from the coal at a much lower temperature than at which they will burn. Therefore, although the temperature of the firebox is lowered below the burning point of the gases, a considerable amount of these gases are driven off from the coal even at this lowered temperature. The result is that the heat which would have been possible to obtain from these gases under normal conditions is lost because the gases are pulled through the flues and out of the stack before they had a chance to burn. On the other hand, had the coal been added in small quantities, the temperature of the firebox would have been maintained at the burning point of the gases.

CONDITION OF COAL, A FACTOR IN COMBUSTION

The condition of coal will determine to some extent how it should be fired.

COARSE (lump) coal will require a somewhat deeper fire than slack coal.

SLACK (fine) coal is burned to a large extent in suspension and a somewhat lighter fire is maintained.

MIXED lump and slack coal requires more frequent inspection of the fire to be sure that an even distribution of the coal is maintained.

WET coal requires watching, especially at the beginning of the trip, to see that it does not stick to the distributing table and that it is being distributed evenly to all portions of the fire.

ASHES

There is a great difference in the kind and amount of ashes produced when coal from different mines of the same seam is being burned. Some coal may produce very little ash; others may produce a large amount. It is reasonable to expect that coal which produces the most ash will produce the least heat per ton burned.

Ashes melt when they are subjected to sufficient heat. When they melt they run together to form clinkers. Ashes from coal high in sulphur and iron content will melt at a much lower temperature than ashes from other kinds of coal. When ashes tend to melt at a low temperature and form clinkers, the grates must be handled carefully to keep the fire clean.

A clean fire will allow a good flow of air through the grates and thus prevent the ashes from reaching the melting point.

It is important to keep the fire clean and special notice should be taken of this point when burning high ash coal. When ashes accumulate on the grates they restrict the flow of air through the fire.

CLINKERS

Some coal has a greater tendency to cake and clinker than others. At any rate, clinkers are caused by improper firing which causes the ashes to melt. Most coal forms a cake clinker before the melting point of the ash is reached. If the clinker is discovered at this point, it can usually be checked and burned out by good crew cooperation and skillful handling of the fire by the fireman.

A clinker in the fire may cause serious difficulty and it may be the cause of a steam failure. Clinkers are treacherous because they start to form without warning, and they grow rapidly after they have started.

It is easier to prevent a clinker from forming than it is to get rid of it after it has started because the measures taken to clear out a clinker may actually make the condition worse.

SUGGESTIONS FOR AVOIDING CLINKERS

Clinkers are formed by ashes (high in sulphur and iron content) which are allowed to get too hot. Heating ashes beyond the clinkering temperature causes them to cake and eventually to melt and run together if the rise in temperature is continued.

Keeping ashes below the clinkering temperature prevents clinkering. The ashes on the grates and below the fire bed are cooled by the air passing through the grates. A thick layer of ashes may start a clinker because the flow of air is restricted.

When a thick layer of ashes is allowed to accumulate, the grates must be shaken excessively to clean the fire and thus the fire bed is disturbed. This action may work partially burned coal down into the ashes. The coal continues to burn and may raise the temperature of the ashes enough to start a clinker.

When a clinkering coal is being fired, the fire bed should not be disturbed with the fire rake as this action might bring points of melted ash together, or mix ash already near the melting point with burning fuel, thus causing a clinker to form.

A bank can form a clinker because the air does not pass freely through that portion of the fire, and when this happens, the ashes under the bank may be heated to the melting temperature.

If a light spot develops in the fire bed, great care must be taken when coal is being added to build up the low area, because a large amount of coal at one time tends to form a clinker. Working fresh coal down into the fire by using a rake or by violent shaking of the grates may also start a clinker.

GENERAL FIRING INFORMATION

Carry as light a fire as possible to maintain the proper boiler pressure. This will depend on how hard the locomotive is being worked. If the fire is too thin, the draft is likely to tear holes in it. The steam pressure will fall if the fire is too thin.

A fire that is too thick may be caused by too much ashes on the grates. A thick fire slows up combustion by restricting the flow of air through the grates. A thick fire causes clinkering and prevents the fly ash from passing out of the stack.

Prepare the fire before reaching a heavy grade so that it will withstand the heavy exhaust. Cut down on the coal supply far enough before the start of a down grade so that the fire will be in good condition for light work and so the safety valve will not raise.

The draft produced by the working locomotive controls the amount of coal that can be burned in that locomotive. Feeding coal beyond this rate results in a thick fire. Feeding coal below this rate results in a drop in steam pressure. To maintain a fire of uniform thickness, more coal must be fed when more steam is being used and less coal must be fed when less steam is being used.

Stop firing far enough from a station to insure gases being consumed before the throttle is closed and thus reduce black smoke and prevent safety valves from opening.

FUEL CONSERVATION "DON'TS"

1. Don't leave the terminal unless the fire is in good condition.
2. Don't feed coal faster than it can be burned.
3. Don't overfill the scoop shovel.
4. Don't allow safety valves to open unnecessarily.
5. Don't use the blower except when necessary.
6. Don't leave the fire door open too long.
7. Don't leave the fire door open while descending grades or stopping.
8. Don't let the firebox temperature go down below the necessary temperature required for perfect combustion.
9. Don't permit the fire to get heavy or dirty.
10. Don't shake the grates excessively.
11. Don't permit banks to form in the firebox.
12. Don't allow the fire to die out in the forward end of the firebox; this may cause flues to leak.
13. Don't allow rock, iron, wood or other foreign matter to be fed into the stoker.
14. Don't throw large lumps into the fire.
15. Don't leave the departure track if the coal in the tank is not properly trimmed.
16. Don't knock coal off by careless handling of tools.
17. Don't allow coal to fall out of the gangways.
18. Don't allow the locomotive deck or apron to get dirty.
19. Don't "die on a full stomach"; never feed more coal than needed.
20. Don't forget to watch the water level in the boiler.

FIRE CONDITIONS

HOW TO PREPARE A FIRE

INTRODUCTION

The importance of starting out with a good fire cannot be over-emphasized. Serious difficulties may be encountered on the road if the fire is not properly prepared and cared for before starting out. The fire should be banked according to local instructions, should be in good condition and the steam should be within fifty pounds of maximum pressure when the locomotive is turned over to the engine crew.

The horseshoe type of bank should be used in preparing the fire for the ready track and it should be in such condition that it would not be necessary to have it disturbed when the locomotive is ready to leave. This type of bank provides a good foundation upon which the fireman can build up his fire. Only enough coal should be added, by hand, to fill in the center of the bank. When adding coal to the fire on the ready track, the blower should be used until the smoke clears up.

HOW TO CARE FOR THE FIRE WHILE THE LOCOMOTIVE IS ON THE READY TRACK

A. Inspect the fire by the following procedure to ascertain that it is properly banked.

NOTE: Follow local instructions for preparing fire.

1. Slightly open blower valve to remove gases from firebox.
2. Open the fire door by depressing the foot pedal or by the hand lever.
3. Inspect the banked fire. The purpose of the banked fire should be judged according to the length of time the locomotive will be on the ready track. The purpose of this bank is to provide a well-coked foundation upon which the fireman can build up the fire before train departure.

NOTE: The bank should not be disturbed while the locomotive is on the ready track. Disturbing the bank will cause the fire to burn out:

4. Watch the boiler pressure and prevent the engine from popping while maintaining the fire on the ready track.

B. Add coal to the banked fire by the following procedures.

1. Turn on the blower to provide the necessary draft to

maintain the fire and to control smoke while coal is being added with the scoop.

2. Use the scoop to add sufficient coal to the bank in the firebox to maintain steam pressure. The amount of coal required will depend on how long the locomotive will remain on the ready track.

3. Build up the fire so that it will be in good condition, and have within fifty pounds of maximum steam pressure when the locomotive is taken over by the crew. The recommended boiler pressure for the locomotive is shown on the "badge plate" located on the back boiler head.

FIREMAN'S DUTY IN PREPARING FIRE

1. Turn on the blower to provide the necessary draft to control smoke when preparing the fire.

2. Do not disturb the horseshoe bank prior to leaving. Add enough coal to build up the boiler pressure to approximately maximum pressure. Before departing see that the fire is well burned through and slightly heavier in the back end than the rest of the fire bed.

NOTE: Although the use of the fire hook tends to start the formation of clinkers, it is permissible to use it at this time because the firebox temperature is low and there is little possibility of clinkers starting to form.

3. Check the coal and water in the tender. Check the water level in the boiler according to local instructions. The water level should not exceed two guages at any time. Refer to section "How to Pump the Boiler."

4. Shut off the blower after the fire is prepared and the smoke is cleared up. The stoker should not be used until the locomotive is departing with train. However, the stoker should be tested at this time to make certain that it is in good operating condition before leaving the terminal.

5. Start stoker operation after leaving the terminal. Adjust the stoker speed to maintain the proper boiler pressure and adjust the stoker jets to distribute the coal evenly over the fire.

NOTE: While the locomotive is on the road, banks in the firebox should be avoided. A clean, even fire should be maintained at all times. For maximum locomotive performance see instructions for firing different kinds of coal on the road.

6. Inspect the fire frequently.

Inspect the fire frequently.

HOW TO INSPECT THE FIRE

INTRODUCTION

Many bad fire conditions can be prevented before they develop into serious difficulties if the fireman forms the habit of frequently inspecting the fire.

There are so many factors which have a tendency to change fire conditions that frequent inspection of the fire is absolutely necessary. Shortening the cut-off, or "easing-off" on the throttle, requires a decreased amount of coal in the firebox and decreases the amount of water required to maintain the boiler water at the proper level. Firebox conditions are changed when the coal changes from lump to slack or from slack to lump, or if the condition of the coal changes from dry to wet or from wet to dry. Jet pressures have an effect on the condition of the fire. The stoker speed and the stoker-jet pressures must be adjusted if the condition of the coal changes or if the load conditions change.

By frequent inspections the fireman can detect the formation of clinkers, and can prevent plugged arches, banks, and holes. Any condition in the firebox which causes unbalanced draft tends to cause a carryover of cinders and unburned coal which may result in a plugged front-end netting. Through frequent inspection most difficulties can be detected and corrected before they become serious.

An experienced fireman will make all of the inspections described in this unit in a very short time and will then take the necessary steps to correct any bad conditions revealed.

PROCEDURE TO BE FOLLOWED FOR INSPECTING THE FIRE

Examine the fire frequently by following the procedures in this unit, so that the exact conditions in the firebox will be known at all times. Frequent inspection gives the fireman a chance to correct troubles before they affect performance.

BEFORE THE DOOR IS OPENED:

1. Observe the pressure indicated on the stoker-engine guage. Knowing the pressure before shutting off the stoker will give the fireman a basis for setting the stoker speed when stoker operation is resumed. After a little experience the fireman will be able to judge the best pressure for different operating conditions.
2. Shut off the stoker by closing the stoker throttle.

CAUTION: The fire door should not be opened for fire inspection until the supply of coal to the firebox is stopped and the gases have been allowed to burn off.

3. Shut off the stoker-jet intermediate valve.
4. Observe the pressure recorded on the boiler-pressure gage, for if the boiler pressure changes it may be necessary to change the stoker-jet pressure to compensate for the change in boiler pressure. A reduced boiler pressure reduces the pressure to stoker engine as well. Watch the stack and the discoloration of smoke for an indication of how fast and efficiently the stoker is feeding the coal to the fire.

INSPECT THE WHOLE FIRE

1. Open the fire door.
2. Take an overall view of the condition of the fire. Experience will enable the fireman to judge the general condition of this fire quickly.
3. Check the whole fire. Maintain the minimum thickness of fire necessary to produce the required amount of steam.
4. Inspect the fire for light spots. Light spots and holes cause a drop in boiler pressure, and the increased draft through that portion of the fire may cause a carry-over of cinders and coal which may plug the front-end nettings.
5. Inspect the condition of the fire in the back end and in the back corners. The back corners are often a source of trouble because the grates slant downward toward the front and the vibration of the locomotive causes the fire to creep off the back grates.
6. Look for high spots, drifts, or banks which are the result of improper distribution and failure to inspect the fire frequently. High spots develop into banks which are one of the common causes of clinkers.

CORRECT BAD CONDITIONS REVEALED BY IN-SPECTION

1. Correct bad fire conditions by following procedures outlined in the unit "How to Correct Bad Fire Conditions."
2. Resume stoker operation. Increase the stoker speed if the boiler pressure should be raised. Reduce the stoker speed if the demand for steam is decreased.
3. Resume the stoker-jet operation. Regulate the jet-manifold pressure to the same pressure that was being used before the fire was inspected unless the condition of the fire indicates that more or less pressure is required to maintain an even depth of fire over the grates.

4. Make frequent inspections to make certain that the fire is in good condition and that any bad conditions that may have been discovered have been corrected.

How to Fire Different Kinds of Coal

INTRODUCTION

The job of firing a locomotive would be a relatively simple procedure if the fireman was assured of having a supply of good quality coal, uniform in size, in proper condition to fire, and running under ideal firing conditions. However, this is not the case. The desirable grade of coal is not always available and conditions constantly vary. Therefore, a fireman must be prepared to fire whatever coal he finds in his tender, and sometimes under unfavorable conditions.

Some firing conditions must be treated individually, but there are some firing practices which apply to all firing situations which, if followed, will save a considerable amount of trouble for the fireman and enable him to get good efficiency from the locomotive he is firing.

STANDARD FIRING PRACTICES

1. Carefully inspect the coal in the tender and at the point of distribution.
2. Maintain an even stoker speed to keep the fire at the minimum depth that will maintain the desired boiler pressure.
3. Adjust the stoker jet pressure to maintain an even distribution of coal over the whole fire. Change jet pressures as necessary if the size, grade, or condition of the coal delivered to the distributing table changes.
4. Inspect the fire frequently and correct unsatisfactory conditions.
5. Hand fire the light spots when necessary to maintain a level fire.
6. Move the grates frequently, but lightly.
7. Maintain correct water level at all times by even pumping.
8. Fire according to the condition under which the locomotive is working.
9. Avoid excessive use of the fire hook.
10. Call the engineer's attention to bad fire condition so that the locomotive can be worked accordingly.

Wet coal, high-slack coal, clinkering coal, and high-ash coal can be fired successfully if the above practices are observed together with the suggestions given below for firing different kinds of coal.

STANDARD FIRING PROCEDURE TO BE FOLLOWED AT ALL TIMES

1. Observe the condition of the coal at the point of delivery. The jet setting will vary according to the size and condition of the coal being burned. Do not allow rock, iron, wood, or other foreign matter to be fed into the stoker if it can be detected in the coal supply and removed before it enters the conveyor.
2. Prepare the fire before starting out as explained in the section "How to Prepare the Fire."
3. Start the stoker after the locomotive is underway according to the instructions in the section "How to Operate a Stoker."
4. Set the stoker jets as explained in the section "How to Adjust Stoker Jets."
5. Adjust the stoker speed to run at an even speed. When small lumps and slack coal are being fed into the stoker, less steam pressure is required to maintain an even supply of coal to the distributing table than when coal containing large lumps is being used.

   Extra power is required to break up lumps of coal by the crusher.

NOTE: An even stoker speed makes possible an accurate setting of the stoker jets.

6. Check the speed of the stoker by observing the amount of coal being delivered to the distributing table. The amount of coal required to keep the fire in proper condition depends on the conditions under which the locomotive is being worked and on the kind of coal being fired. Close observation of the conditions in the firebox and the conditions under which the locomotive is working will enable the fireman to make adjustments, when they are necessary, and get the best efficiency out of every ton of coal burned.
7. Keep a constant check on the gages. Changes in reading tell a story which enables the fireman to take steps to correct conditions that affect the fire before they become serious.

   a. BOILER PRESSUREGAGE — A change in boiler pressure may require a change in the volume of coal necessary to maintain the correct boiler pressure, and an adjustment of the stoker jets to maintain an even distribution of the coal. A change in boiler pressure may require attention to the fire to bring the pressure up, if it is dropping, or may require a reduction in the amount of coal being supplied to the fire to avoid popping.

b. STOKER ENGINE GAGE—Fluctuation of this gage gives an indication of what is happening in the tender conveyor unit and the crusher. The gage is also an indication of the stoker engine speed. If the gage reading goes up approximately to boiler pressure and stays there, it is an indication that the stoker has stopped. The cause for a stoker engine to stop may be lack of lubrication or some obstruction being lodged in the crusher or conveyor.

c. STOKER DISTRIBUTING JET GAGES — A study of these gages will make it possible to set jets for even distribution of the coal to all parts of the firebox. A frequent check of the coal being fired, to detect a change in the size or condition, will make it possible to adjust jet pressures before the fire is affected. When the jets are to be shut off, note the pressure shown on the gages so that jet operation may be resumed with the same pressure if desired.

d. WATER GLASSES AND GAGE COCKS—Low water can cause serious damage to the crown sheet and may endanger life. High water spoils the Superheated Effect of the locomotive and Destroys Lubrication. The water level in the boiler should not exceed two gages at any time.

8. Water should be added to the boiler at an even rate to maintain two gages of water as indicated by the water glasses and gage cocks. See section "How to Pump the Boiler" for instructions.

9. "Move" the grates often to prevent accumulation of ashes on the grates, to avoid clinkers, to facilitate an adequate supply of air and to maintain the proper fire depth.

10. Inspect the fire often. The frequency of the inspection will depend on the kind of coal being used and how the locomotive is being used. A bad condition can be corrected before it causes any serious trouble if it is discovered soon enough. See section "How to Inspect the Fire."

11. Adjust stoker speed, stoker jet pressures, or vanes as required to improve the fire and thereby maintain an even boiler pressure. See section "How to Correct Bad Fires" for instructions.

SUGGESTIONS FOR FIRING WET COAL
Follow the standard firing procedure outlined above and

apply the following suggestions.

1. Inspect the fire frequently. When the coal is wet, the even distribution of the coal over the fire is more likely to be affected because of the tendency of the wet coal to stick together or to stick to to the large pieces.
2. Adjust the jets, if necessary, to slightly higher than normal pressures as the wet coal tends to stick to the distributing table. This condition is more noticeable at the beginning of the trip before the distributing table is heated thoroughly. Use the slide hook, if necessary to clean the distributing table, until it is heated thoroughly.
3. Use the jet hook or poke-out rod to dislodge the coal from the distributing table, even after the jet pressures have been increased. Wet coal has a tendency to bake on the distributing table if not removed. If it is allowed to accumulate, the increased jet pressure may blow the coal back under the arch.
4. Inspect the fire after each adjustment of jet pressures to make certain that the coal is being distributed evenly.

SUGGESTIONS FOR FIRING CLINKERING COAL

Follow the standard firing procedure outlined in the first part of this section and apply the following suggestions:

1. Adjust the stoker speed to AVOID CROWDING the fire. Crowding the fire increases the tendency to produce clinkers and causes excess smoke.
2. Adjust the stoker jets to get an EVEN DISTRIBU-TION of coal and prevent banks from forming. Banks shut off the air through the thick portion of the fire and thus a clinker forms and grows rapidly.
3. AVOID the USE of the FIRE HOOK with clinkering coal. Use it only in emergencies as the hook causes the clinkering substances to run together and form a clinker.
4. Move, or "wiggle" the grates often to insure a free flow of air through the fire. DO NOT SHAKE GRATES VIOLENTLY.
5. Inspect the fire often. It is easier to CORRECT A CLINKERINNG CONDITION BEFORE IT BE-COMES SERIOUS than it is to get rid of a bad clinker.

SUGGESTIONS FOR FIRING HIGH SLACK (FINE COAL

It is an important thing to remember that slack coal is just as good as lump coal and costs just as much. Slack has the same amount of gas, coke, ash, and heat value; the only difference is that the pieces are smaller and the feeding, distribution and jet pressures must be watched more carefully.

Follow the standard firing procedure outlined in the first part of this section and apply the following suggestions:

1. Adjust the jets, and vanes if necessary to control the distribution of the coal. The stoker screw tends to deliver more fine coal to the right and more lump coal to the left, when the two are mixed.

2. Carry a slightly heavier fire at the back portion of the firebox. It is good practice to use the scoop to build up the back portion of the fire.

3. Readjust the jet pressures if the coal changes from lump to fine or from fine to lump. Note: Fine coal requires less pressure than lump coal. Excessive pressure will pile the coal in front of the firebox and may result in a plugged arch.

4. Watch all signs of impending trouble such as the color of exhaust from the stack, the gages, the color and action of the flame, the flow of coal at the crusher, the delivery of coal at the distributing table and not the least, be alert for signs of stoker defects. These signs will give warning of trouble in time to make necessary adjustments and thereby prevent the difficulties that are sure to follow if the proper steps are not taken.

5. Know the condition of the fire at all times, through frequent observation, so that adjustments can be made intelligently.

NOTE: Two causes create more trouble for firemen than any others.

(a) Poor distribution of coal, very often due to failure to make adjustments in time.

(b) Feeding more coal to the firebox than the engine will burn.

When the steam pressure begins to drop, it is a natural tendency to increase the feed of coal. DON'T DO IT! Examine the fire first. It may be due to a pile or a hole in the fire, in which case, increasing the feed will only make matters worse. What may be needed is better distribution and this can be obtained by adjusting jets, or feed, and often by adding a scoop of coal by hand. DO NOT increase the feed unless you know by observation that the fire is too light.

INTRODUCTION

In starting out, the fireman should endeavor to have a

light, level, bright fire. These three conditions should always exist in a firebox — the thickness of the fire should be regulated by the class of fuel, the drafting of the engine, and the weight and schedule of the train.

It is further recognized that each class of locomotive requires different handling on account of different draft and varying coal conditions. However, besides these known variables, there are many other factors which will affect a fire. Fortunately, most of these factors give the fireman a warning before they become serious. It is, however, the responsibility of the fireman to be able to recognize conditions that will affect the fire and know what steps must be taken to correct a bad fire condition before it becomes serious.

The purpose of this unit is to describe some of the conditions that will affect a fire so that the fireman, knowing what may happen to a fire as a result of prevailing conditions, can anticipate bad fire situations before they develop or discover them immediately after they have developed and take proper steps to prevent them from becoming serious.

CHANGES IN KIND OF COAL

If the coal goes from lump to slack, the jet pressures must be DECREASED, otherwise a bank may be formed under the arch.

If the coal goes from slack to lump, the jet pressures should be INCREASED, otherwise a bank may form in the back of the firebox.

If the coal goes from dry to wet or from wet to dry, the jet pressures may require adjustment.

The fireman will know when the changes take place if:

a. He watches the coal being supplied to the stoker.
b. He watches the coal being delivered to the distributing table.
c. He inspects the fire frequently to see that the coal is being evenly distributed.

IMPROPER JET SETTING

If the stoker jets are IMPROPERLY SET banks will form in the firebox. The fire should be built up properly at the terminal and then careful attention should be given to the jet pressures to maintain an even distribution of coal after leaving the terminal and after the stoker is started.

The fireman can detect improper jet pressures if:

a. He watches the grade of coal in the tender and at the distributing tables to detect changes in the kind of coal being fired.

b. He watches the amount of coal being delivered by the stoker.
c. He inspects the fire frequently and watches for banks.
d. He watches the boiler-pressure gauge for an increase or decrease in pressure.
e. When one jet is adjusted the manifold pressure must be reset, as each time a jet is reduced or increased in pressure the manifold pressure varies accordingly.

IMPROPER SUPPLY OF COAL TO FIREBOX

TOO LITTLE COAL will let the fire become light in some portions of the firebox and may result in an unbalanced draft through the firebox. This may cause an excessive carryover of cinders and unburned coal which may clog the netting. A thin fire has a tendency to creep off the back grates. The steam pressure will drop if the fire is too light for the conditions under which the locomotive is working.

TOO MUCH COAL will cause a heavy fire which may result in an accumulation of ash and cinders. Excessive smoke and a waste of coal result from feeding coal to a fire faster than it can be burned.

The fireman can supply the proper amount of coal to the firebox if:
a. He inspects the fire frequently and increases or decreases the supply of coal by regulating the stoker speed as required.
b. He watches the steam pressure and observes the condition under which the locomotive is being worked and then controls the coal supply accordingly.
c. He knows the "run" and can plan ahead to supply coal according to speed demands.
d. He has the cooperation of the engineer who notifies him beforehand of changes in running conditions that will affect the demand for steam.

CUT-OFF AND THROTTLE CHANGES

If the cut-off is shortened or the throttle is eased off, less steam is required and there is a possibility of getting too much COAL IN THE FIREBOX and too much WATER IN BOILER unless the stoker speeds, the stoker jet pressures, and the water supply are regulated to decrease the supply of coal and water according to the reduced demands.

If the cut-off is lengthened, or the throttle is widened, more steam is required and more COAL AND WATER must be supplied. The stoker speeds, the stoker jet pressures, and the water supply must be increased accordingly.

The fireman will know when these changes take place if:

a. There is close cooperation between the fireman and the engineer.

b. He observes the condition under which the locomotive is being worked and anticipates a need for more or less coal.

c. The steam pressure gauge is watched for a rise or fall in pressure.

FIRE WORKS AHEAD ON THE GRATE

There is a tendency for the fire to creep ahead on the grate when there is a change in speed or when the vibration is set up by the engine riding rigid at high speed in short-running cut-off. This causes the fire to become light on the back grate.

The fireman can avoid this by:

a. Inspecting the fire for this condition when the locomotive is riding rough or vibrating, and when the change in speed takes place.

b. Notifying the engineer if the fire condition is serious in order that he may ease off or, if necessary close the throttle to decrease the pull on the fire.

c. Adjusting the jet to maintain an even fire.

d. Shaking the front grate, if necessary, to equalize the draft through the fire before building up the light spots with the scoop.

e. Using the scoop, if necessary to assist in building the fire up along the back.

SLIPPING OF THE LOCOMOTIVE DRIVING WHEELS

Should the locomotive driving wheels happen to slip severely, the momentary excessive draft on the fire caused by this condition may tear holes in the fire. It is essential that the fire is inspected after such a situation has occurred. The fireman should be able to judge how much damage was caused to the fire by the occurrence of this situation and take the proper steps to correct it. If holes are torn in the fire, the condition may be corrected most satisfactorily by adding coal, by hand, with the scoop.

How to Correct Bad Fire Conditions

INTRODUCTION

Bad fire conditions should not be a common occurrence and they can be avoided if the fireman will observe the rules of good firing. However, it is known that they do occur and in spite of proper care and all precautions, unsatisfactory fire conditions develop. Therefore, it is necessary for the fireman to be able to recognize such conditions and know how to correct them.

The occurrence of many of the bad fire conditions is quite common; others are a result of an unusual situation. Regardless of the causes, bad fire conditions may result in serious delays, and even tie up an entire section unless they are discovered soon enough and corrected. This unit is intended to point out some of the bad fire conditions and offer procedures for overcoming them.

HOW TO CORRECT BANKS IN THE BACK OF THE FIREBOX

1. Examine the fire carefully to determine the size and location of the bank. Refer to the unit "How to Inspect the Fire," for information on fire conditions.

NOTE: Banks, as a general rule, are the result of improper distribution and failure to inspect the fire frequently. They are one of the common causes of clinkers.

2. Determine the cause of the bank.

NOTE: The bank may be caused by improper jet pressure, improper adjustment of vanes, the formation of a clinker, or by crowding the fire.

3. Correct the bank by doing any or all of the following:

a. Place the shaker rod on the shaker lever and move the grate under the bank lightly to equalize the flow of air through the fire, and thus prevent the formation of clinkers.

NOTE: The fire hook may be used to level off the bank if care is taken not to disturb the fire bed.

b. Readjust the vanes (on stoker so equipped) by loosening the lock nuts on the vane-control screws and by turning the screws to the right to reduce the amount of coal delivered to the back corners.

c. Increase the steam pressure to the right back and left back distributor jets by opening the jet valves slightly so that the coal will be distributed evenly over the entire fire bed. For information on how to adjust stoker jets refer to the unit "Operating The Stoker."

4. Examine the fire often to check the results of these adjustments, thus making certain that coal is being evenly distributed.

HOW TO CORRECT BANKS IN THE FRONT OF THE FIREBOX

1. Examine the fire carefully to determine the size and

location of the bank as a basis for deciding upon the corrective measures to be taken.

2. Move the grates under the bank slightly.
3. Decrease the pressure on the right front and left front distributor jets by turning the jet valves clockwise.
4. Inspect the fire frequently to make certain that the coal is being evenly distributed.

NOTE: When wet coal is being fired, the jet pressures are sometimes increased so high that the wet coal is blown under the arch. When this happens, the arch may be plugged.

HOW TO CORRECT A PLUGGED ARCH
1. Shake the front grates lightly until the excess ash has been removed.
2. Level off the fire with the fire hook.

NOTE: Use the hook with care to avoid mixing the unburned coal with the ashes and thereby providing a possibility for developing clinkers.

3. Decrease the jet pressures by turning the right front and left front jet valves clockwise to decrease the amount of coal supplied to the front of the firebox.
4. Examine the fire frequently and make certain that the bad fire condition has been corrected and that the coal is being evenly distributed.

HOW TO CORRECT LIGHT SPOTS UNDER
THE DISTRIBUTING TABLE (HT Stoker)
1. Examine the fire carefully to determine the size and location of the light spot.

NOTE: A light spot under the distributing table is one of the most difficult conditions to detect because it is not possible to see directly under the table.

If the boiler pressure drops and the fire directly ahead of the distributing table has an especially bright appearance, it is reasonably safe to assume that there is a light spot or a bare spot under the distributing table.

2. Hand fire the light spot with the scoop, or invert the scoop and place it over the top of the distributing table and close the firedoor. This will deflect the coal down below the table.

CAUTION: If the inverted scoop method is used, close the firedoor to prevent the coal from being blown out of the door.

3. Examine the fire often to check on proper depth and even distribution.

HOW TO CORRECT LIGHT SPOTS ON
THE BACK GRATES

1. Check the fire bed carefully for light spots or bare spots and note the size and location of them.

NOTE: Bare spots occur often on the back grates because the vibration of the locomotive causes the fire to creep ahead of the grates which are on an angle.

2. Call the attention of the engineer to the bad fire condition so that he may "ease off" or, if necessary, close the throttle so that the pull on the fire will be decreased, thereby eliminating the possibility of plugging the netting while the fire is being built up.
3. Hand fire the bare spots with the scoop.
4. Adjust the stoker jet to supply more coal to the back portion by increasing or decreasing the pressure on the right-back and left-back jet.

HOW TO CORRECT FIRE DEPTH—
FIRE TOO DEEP

1. Inspect the fire carefully to determine the depth by estimating the distance between the top of the fire and the distributing table.
2. Reduce the speed of the stoker by turning the stoker-throttle valve clockwise so that the supply of coal will be decreased. The fuel bed should be kept level and just enough coal fed to the fire to maintain the correct boiler pressure. Excess fuel is waste, produces smoke, and may cause clinkers.

CAUTION: Changing the speed of the stoker engine will necessitate a readjustment of stoker jets.

3. Observe the amount of coal being delivered to the distributing table by opening the peep holes in the top of the elevator pipe or discharge box.
4. Watch the boiler pressure.
5. Reduce the fire depth if necessary, by placing the shaker bar on the front grate lever and by rocking the grate carefully. For information on shaking grates, refer to the unit "How to Shake Grates."

NOTE: Do not rock grates so violently that the entire fire bed will be disturbed.

6. Place the shaker bar on the rear grate lever and rock the rear grates carefully until the proper depth has been obtained. It may be well to wait a short time after shaking the front grate before shaking the rear grates and thus avoid disturbing the whole fire at one time.

HOW TO CORRECT A CLINKERED FIRE

1. Inspect the fire to determine the size and location of the clinker. The size of the clinker may be determined by the size of the bank or dead spot and also by the area of the fire which moves when the grates are rocked. Place the shaker rod on the grate lever and open the grates under the clinker enough to break the clinker. This will allow air to pass through the clinker area. If this does not eliminate the clinker, apply the following suggestions as a last resort:
   a. Pull the clinker to the back of the firebox, under the distributing table with a fire hook.
   b. Knock the clinker through the grate if possible.

NOTE: If the clinker cannot be broken up small enough so that it will go through the grates, it should be removed through the fire door.

CAUTION: Use care in handling hot clinkers. The hose may be used to cool the clinker after removing it through the fire door.

2. Hand fire the light spot with a scoop.
3. Readjust the stoker jets to equalize the distribution of the coal so that banks will not form. Banks develop into clinkers.
4. Examine the fire frequently. Frequent inspections are the fireman's best insurance against serious troubles with his fire. A bad fire condition, if it is discovered soon enough, can usually be corrected by making a minor adjustment.

NOTE: If the coal being fired tends to form clinkers, pay particular attention to the following standard firing rules: (a) Keep a balanced draft through all portions of the fire by maintaining an even fire, (b) Avoid overcrowding the fire, (c) Do not allow banks to form, (d) Move the grate often and gently to keep the fire, light and clean, (e) Avoid the use of the fire hook except in emergencies.

SUGGESTIONS ON FIRING TO AVOID DAMAGE TO THE FIREBOX AND TUBES

1. The fire should be heaviest at the sides and the back corners of the firebox to make certain that holes do not develop at these points. Holes here would cause cold air to strike the side sheets or the back sheets of the firebox.
2. Avoid too much coal in back of the firebox as this causes burned distributing tables and clinkered fire.

3. Maintain an even steam pressure, and pump an even water level to avoid too rapid expansion or contraction of firebox sheets and boiler flues.
4. Watch the water gages closely to avoid low water. This condition is dangerous and may cause burned crown sheet.
5. Avoid holes, light spots, or banks in the fire bed, as these cause an uneven draft through the firebox.

THE STOKER

A cutout showing the installation of a Standard stoker, Type "HT", on a modern locomotive. On Reading Company freight power the stoker engine is located beneath the cab on the fireman's side of the locomotive.

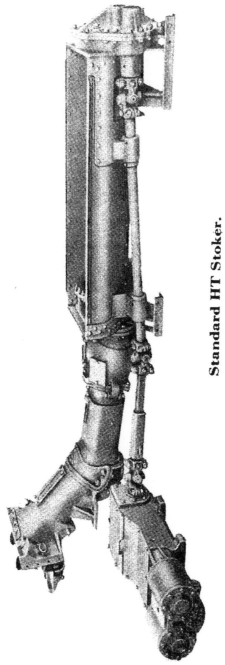

Standard HT Stoker.

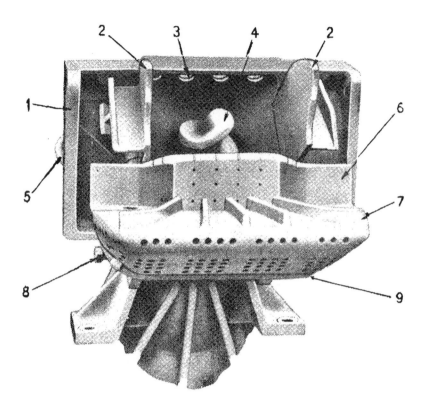

**Front View of HT Stoker showing location of Elevator
Screw, Vanes, Jet Plate and Distributing Table
in relation to the Elevator Pipe.**

1. Elevator Pipe
2. Right and Left Vanes
3. Peep Holes
4. Elevator Screw

5. Vane Adjusting Handles
6. Jet Plate
7. Distributing Table
8. Jet Plate Set Screw

9. Air Sealing Apron

THE STOKER

GENERAL DESCRIPTION

Primarily, the stoker is a simple machine for conveying the coal from the tender to the firebox and distributing it uniformly over the fire. But, there are several important units in a locomotive stoker.

The stoker has five principal parts: the stoker engine, the tender conveyor unit, the intermediate conveyor unit, the elevator unit, and the distributing unit. Most of the stokers operate on the same principle.

Although it may not be necessary for a fireman to know all the individual parts which make up the stoker, it is not only desirable but very important that he has a thorough knowledge of how the system operates. This information alone will be a big help in his effort to get the most steam out of the least amount of coal and physical effort and will naturally result in a better job of firing the locomotive.

THE STOKER ENGINE

The location of the stoker engine is optional. Some units are installed under the cab of the locomotive, others are on the tender. However, the stoker engine is the unit which supplies the mechanical power for operating the conveyor units of the stoker.

The stoker engine is designed to operate on saturated steam. The power is transmitted through a shaft and universal joints to the gear box.

The speed of the engine can be controlled by a throttle valve located in front of the fireman's position in the cab. Naturally the speed of the stoker engine will determine the amount of coal that is delivered to the distributing table of the firebox.

The engine is reversible so that in case a large lump of coal or other obstruction should clog up the conveyor, the direction of the conveyor screw can be reversed and the obstruction removed.

THE TENDER CONVEYOR UNIT

Directly underneath the coal doors of the tender, provisions have been made to mount the tender conveyor unit in such a way that practically all the coal in the tender can be fed to the conveyor and through to the firebox. The unit consists of a trough, conveyor screw, crusher, and gear box. The length and width of the trough may vary to suit the capacity of the tender.

The reduction gears of the gear box provide the revolving motion to the conveyor screw. This conveyor screw carries the coal forward to the crusher where it is broken up to the proper size necessary to provide an even flow for satisfactory firing.

THE INTERMEDIATE UNIT

The intermediate unit consists of an intermediate conveyor screw inside of two telescoping tubes and provides a connection between the tender conveyor unit and the elevating unit. It is located below the deck of the cab.

A ball and socket connection between the tender conveyor unit and the intermediate unit and again between the intermediate unit and the elevating unit provides the flexibility necessary between the locomotive and the tender. The intermediate conveyor screw is connected to the tender and elevator conveyor screws by universal joints to provide flexibility between the locomotive and the tender and assure a smooth action of the conveyor screws.

THE ELEVATOR UNIT

The elevator unit consists of an elevator pipe and an elevator conveyor screw. They are connected to the intermediate conveyor unit by a ball and socket joint and a universal joint. These connections are located below the deck of the cab. The upper end of the elevator pipe is connected to the distributing unit. As the name would imply, the elevator unit raises the crushed coal (from the intermediate conveyor unit) to the distributing unit.

THE DISTRIBUTING TABLE

The distributing unit consists of the distributing table, the jet plate and vanes. The distributing table, which is fastened to the jet plate, has deflecting ribs on the top. The distributing table is protected from over-heating by air, deflected against the underside of an air-sealing apron. The jet plate is provided with a series of holes through which steam is blown to distribute the coal over the fire bed.

OPERATING THE STOKER

INTRODUCTION

A fireman's success depends on knowing what to do and also what not to do. The best fireman is certainly not the man who can put the most coal into the firebox in a given time. But a good fireman knows just how much coal is required to do the given job. He knows how to use the coal to the best advantage. He knows how to use all the heat units of the fuel to the highest possible degree, and this certainly does not mean using more coal. On the contrary, it means using less — a great deal less coal. Therefore, a fireman must be sure to convey NO MORE COAL to the firebox than can be burned properly. He should try to carry a thin, level, white fire and keep a clear stack.

An-ideal firing condition is one in which coal is fed into the firebox continuously, at a uniform rate, in accordance with the demand of the locomotive. The coal supply is then as uniform as the air supply which together will give correct combustion and make the most efficient firing condition. It is easy to understand why it is more economical to continuously feed a small amount of coal over the entire fire than it is to feed a large quantity at spaced intervals. In the first place, it results in better combustion and in the second place, it does not chill the fire with a large quantity of green coal.

To help the fireman get these results, with the least amount of effort, the mechanical stokers are so designed that any amount of coal necessary may be conveyed automatically to the distributing table and spread evenly over the fire.

Although the stoker will convey the coal from the tender to the distributing table and spread it evenly over the whole fire, it is not automatic. The efficient operation of the stoker depends on the skill and intelligence of the fireman. The fireman has to stay right on the job and watch his fire very closely. Various conditions will make it necessary to either change the speed of the stoker engine or readjust the jet pressure or perhaps remove some obstruction from the conveyor unit. It is the fireman's duty to recognize these varying conditions and be in a position to make the necessary adjustments. For example, stoker speeds should be changed as the size, grade, or condition of the coal varies, and when changes in the locomotive speed or train load create a demand for more or less steam. The good fireman develops a habit of observing the stoker gauges and steam gauges frequently and also of checking the coal supply through the peepholes.

BEFORE STARTING THE STOKER

1. Check the grade, size and condition of the coal. The kind of coal will give you some indication of the stoker's speed and stoker jet pressure which will be required. Less pressure is required for slack coal than for lump coal.
2. Local conditions governing the start make it impossible to lay down a fixed rule governing the kind of

fire that must be on the grates. However, the fire should be bright and level over the entire surface of the firebox and sufficiently heavy to withstand the additional draft and jar of starting the locomotive.

3. If it is found necessary to build up the fire before starting the locomotive, this should be done by hand. The stoker should not be used to build up a fire at the terminal.

STARTING THE STOKER ENGINE

1. Open turret valves admitting steam to engine and jet lines. Next gradually open main jet valve, then open the five jet valves separately to see that jet holes are free from obstruction. Open stoker engine valve slowly to permit any condensate to escape through automatic drain valve. Place operating lever in forward porition and adjust engine valve to run the stoker at desired speed.

2. Set the reversing lever in the forward position. his lever operates the stoker engine reverse valve. The location of the reversing lever will be dependent upon the location of the stoker engine. On installations where the stoker engine is under the cab of the locomotive, the reversing lever will be on the left side of the cab in front of the fireman's seat. When the stoker engine is in the tender, the reversing valve lever is on the left front leg of the tender.

Note: There are two different types of reversing valves and they work distinctly opposite to each other. For example, on the HT and HT-1 stokers, the lever should be DOWN for the forward position and all the way up for the reversing position, while on the HT-2 stoker the lever should be UP for the forward position and all the way down for the reverse position. However, on both types of stoker engine designs, the neutral or safety position is in the middle.

3. Open the stoker engine throttle valve slowly at first, to allow the cylinders to heat up and permit any condensation to escape through the automatic drain, then adjust the valve to admit the amount of steam necessary to develop the stoker engine speed desired. The desired speed of the stoker engine is determined by the size, grade, and condition of the coal as well as the load which the train carries, the grade it has to travel, and the operating conditions of the locomotive.

Left side of cab of Class T-1 locomotive showing stoker and jet operating valves and gauges.

OPERATING THE STOKER

1. The stoker engine should be operated with steam through the regular throttle valve. However, when it it necessary to increase the steam pressure to the stoker engine rapidly in order to crush an exceptionally large or hard lump of coal, the BOOSTER valve should be opened. But, as soon as possible, the booster valve should be closed and the stoker engine should be continued to operate with steam through the regular throttle valve.

2. Pull the first slide plate in the tank forward, with a hook, to admit the coal to the conveyor. The slide plate should not be pulled forward the full length but just enough to feed coal at the proper rate to the conveyor. A slide opened only about half way will provide enough space to feed the coal through. However, when using lump coal, the slide plate will have to be opened much wider.

3. Check the amount of coal being delivered to the distributing table and adjust the five jet valves so as to get an even distribution of coal over the entire grate area.

   Do not feed too much coal—carry a light fire. In firing with a stoker the fire should be carried considerably lighter than in hand firing.

   Run the stoker as slow as possible, feeding just enough coal to supply the fire for the work being done by the locomotive. On extra light runs, the locomotive will not put a heavy demand on the stoker and it will frequently have to be shut down part of the time.

   Close observations of conditions in the firebox and of the conditions under which the locomotive is working will enable the fireman to make adjustments, if necessary, and get the best efficiency out of every ton of coal burned.

4. Should the steam pressure begin to drop, most fireman almost instinctively increase the stoker engine speed so as to deliver more coal to the firebox. As a matter of fact, nine times out of ten the trouble is too much coal on the fire—more coal than can be burned, banks, holes in the fire, unequal distribution or some other cause. Before increasing the stoker engine speed, check the fire to determine the cause for the drop in pressure.

OPERATING THE JETS

Correct stoker speed and stoker jet pressure are the two most important factors in successful stoker firing. The fire-

Correct stoker speed and stoker jet pressures are the two
most important factors in successful stoker firing.

man must understand that the amount of coal delivered to the distributing table is controlled by the stoker speed, and the distribution of the coal evenly over the entire grate area is dependent upon proper adjustment of jet pressure.

When the fireman notices that the stoker is overcrowding the fire, he can avoid serious trouble by reducing the speed of the stoker engine so that it delivers only enough coal to maintain boiler pressure. There are other factors that require readjustment of stoker jet pressure, such as wet coal, dry coal, slack coal, and lump coal.

Inspect the fire frequently. Many bad fire conditions can be corrected before they develop into serious difficulties providing the fireman forms the habit of watching his fire frequently and compares it with the demands of the locomotive.

1. Turn on the intermediate jet valves slowly. If the intermediate jet valves and the five jet valves are opened suddenly to full pressure it would damage the jet plate.

2. Adjust the upper left-hand valve marked left back. This valve controls the amount of coal supplied to the left back corner. The difficulty of supplying the back corners with an adequate amount of fuel causes these places to be troublesome.

3. Adjust the upper right-hand valve marked right back to distribute the coal to the right back corner which is often likely to be short of coal. The right back and the left back jets should be only given enough steam to remove the coal from the corners of the distributing table.

4. Adjust the valves marked left front and right front which control the jets that distribute the coal to both sides and middle of the fire bed.
   A duplex steam gauge with a red hand for the left front and a black hand for the right front section shows the pressure being used on these jets.

5. Adjust the center valve. This valve is called the fine coal jet line and it controls the distribution of fine coal over the front and center portions of the firebox, through holes located below the jets for the front corners. A separate steam gauge shows the pressure and, as these jets distribute fine coal, only the minimum pressure (approximately 15 to 25 lbs.) should be used.

6. The right and left vanes are adjustable by means of hand screws. They control the amount of coal delivered to the back corners. After obtaining proper distribution, it should not be necessary to change their setting.

7. Observe the reading on the boiler pressure gauge as any changes in boiler pressure should be compensated for by adjusting the intermediate jet valve to make an even distribution of coal in the firebox.

Note: Upon the arrival at the engine terminal, place the stoker operating lever in neutral position and close the stoker engine throttle valves and the jet valves. The intermediate jet valve is provided with a bleed hole. This precaution prevents overheating the distributing table.

DESCRIPTION OF DIFFERENT KINDS OF STOKERS

There are four different types of stokers in use on the locomotives of the Reading Railroad but all of them work on the same basic principle. The distribution of coal is accomplished by means of jets of steam which can be regulated by the fireman so that the coal is evenly distributed to all portions ofthe grate area. The volume of coal delivered by the different types of stokers is controlled by the speed of the stoker engine. The general operations are very similar on all types of stokers and the function of all mechanical stokers is to provide the firebox with coal in the proper quantity and to distribute it evenly to all parts of the fire bed.

The use of mechanical stokers for locomotives has proved conclusively that stoker firing results in greater hauling capacity at a higher rate of speed than is obtainable from hand firing. The fireman who understands the operation of the stoker and can make use of the advantages they offer will relieve himself of a great deal of physical effort and be able to do a better firing job.

Recognizing that there is some variation in the operating mechanisms and controls, this unit is intended to point out some of the distinctive features of the different kinds of stokers so that the fireman may be in a better position to take advantage of them.

THE H-T STANDARD STOKER

The H-T Standard stoker is built on the principle of the screw-to-jet feature. The coal is conveyed from the locomotive tender to the distributing table where it is distributed by a jet blast. The stoker is composed of five principal units:

a. Tender conveyor unit, located beneath the coal supply.
b. A telescopic intermediate unit, situated between the tender and locomotive.
c. An elevator pipe, which feeds the coal to the locomotive firebox.
d. A distributing unit, fastened to the elevator pipe which

includes a jet plate, distributing table, and air sealing apron.

e. A 7 x 7 engine which provides the necessary power to drive the conveyor mechanisms.

THE CONVEYING UNITS

1. Coal is delivered to the stoker, into the tender conveyor unit, and carried forward by the tender intermediate and elevator screws and delivered to the distributing table. The tender conveyor unit is equipped with a CRUSHER located at the forward end of the tender trough. This breaks the larger lumps of coal to a proper fire size and permits the smaller lumps to be carried through the crushing zone without further breaking.

2. The intermediate unit, as used on the HT standard stoker, consists of two telescope tubes, one end flexibly connected to the tender trough, the other to the elevator pipe by ball and socket joints. Within these telescopic tubes, a screw is also flexibly connected to the tender and elevator screws.

3. The elevator pipe and screw are connected with the intermediate unit and screw just beneath the cab deck. From this point the elevator pipe extends upward through the deck and its upper end conforms to the shape of the lower half of the fire door opening. The jet plate fits into and forms the bottom surface of the upper portion of the elevator pipe. It has sixteen drilled holes of appropriate diameter and at different angles. These are divided into five groups serving definite sections of the firebox.

   1. Elevator Pipe
   2. Right and Left Vanes
   3. Peep Holes
   4. Elevator Screw
   5. Vane Adjusting Handle
   6. Jet Plate
   7. Distributing Table
   8. Jet Plate Set Screw
   9. Air Sealing Apron

4. The distributing table is hooked onto the jet plate. It has deflecting ribs on the top surface. These, together with the jet plate and adjustable vanes, provide for sufficient flexibility to distribute coal to any part of the firebox.

5. The stoker is driven by a double acting, two cylinder, variable speed, reversible steam engine, The engine

may be reversed by a lever which operates the valve in the piping. This changes the direction of the steam to the intake and exhaust ports of the engine. The reverse valve used to control the direction of the stoker engine is an ordinary piston type valve operated by means of a handle and shuttle lever connections. It has three different positions: forward, when the handle is all the way down; reverse, when the handle is all the way up; and neutral or safety position, when the handle is central.

IMPORTANT

When removing obstructions from Conveyor system, BE SURE the conveyor reverse lever is in the safety (neutral) position and that the steam to the stoker engine cylinders is shut off. This means that both the operating valve and the booster valve must be closed.

FAILURE IN SERVICE

When Duplex Type D-1 and D-2 stokers stop or fail to operate in service, follow this procedure to determine causes:

1. Move operating valve lever to its lowest position.
2. Place tender conveyor reverse lever in central position.
3. Place right elevator pawl shifter in neutral position.
4. Move operating valve to center or running position.

If left elevator does not operate now, the obstruction is in the left elevator.

If the left elevator does operate, cut in the right elevator by lowering pawl shifter. If it operates, obstruction is in the tender conveyor.

To determine whether trouble is in stoker engine, cut out tender conveyor by placing reverse lever in central or neutral position and raise both pawl shifters to neutral position. If stoker engine does not operate, trouble is in stoker engine.

TO REVERSE CONVEYOR SCREW

Lower operating valve lever on boiler head to bottom position. Move screw conveyor reverse lever back to rear position. Raise operating valve lever on boiler head to center position. Conveyor screw should now run in reverse motion.

TO REVERSE RIGHT AND LEFT ELEVATOR SCREWS

Raise elevator pawl shifter on top of vertical shaft to the top position. Before reversing elevator screw, stop conveyor screws as outlined or stoker will be blocked with coal.

When it is noted that an unusual amount of steam is required to operate stoker, check oil feed. It is important that lubricator feeds 2 drops of oil per minute to stoker engine while it is in operation.

It must be understood that with stoker reversing rod broken, stoker can be operated by hand by using operating valve lever on the rear boiler head.

With the stoker engine properly lubricated, and stoker engine does not operate with both pawl shifters in neutral position and tender conveyor cut out, do not delay further, but proceed promptly by hand firing.

STANDARD STOKERS (INCLUDING DUPONT SIMPLEX TYPES)

The speed of these stokers can be determined by the vibration of the pointer on the steam gauge and by the movement of the conveyor screw. When the stoker steam gauge pointer goes up to practically boiler pressure and remains stationary the stoker has stopped. This indicates that an obstruction exists somewhere in the conveyor system.

To determien cause, reverse stoker a few times, using the booster valve if necessary. If stoker then does not operate and only makes a few turns, indications are that conveyor screw is obstructed by a piece of wood, or other foreign substance that would not pass through the crusher. It might also be caused by baked coal in the pot (horizontal conveyor) or one of the forward conveyor screws being broken.

Flush the tender and conveyor trough with water from the spray hose and continue to reverse stoker to loosen the obstruction in the crusher. If no obstruction is found at crusher, check coal in pot and if carboned (baked), loosen with hook and open steam valve in the pot to clean out line. If this does not correct condition, apply water from spray hose. This should correct the trouble.

When the locomotive is going to be at rest for any length of time on line of road or when it is turned in at end of run, the stoker should be run in reverse a sufficient length of time to remove coal from the vertical conveyor. This will help to overcome baking coal in pot.

Part No.	Parts Per Stoker	Name of Part	Part No.	Parts Per Stoker	Name of Part	Part No.	Parts Per Stoker	Name of Part
*912(v)	1	Main Steam Gauge—Single Pointer	*1153(v)	1	Tender Conveyor Screw	28086-A	1	Elevator Pipe Door
*1027(v)	1	Distributor Jet Manifold, Complete	*1154-A(v)	1	Tender Trough	28104-A	2	Universal Joint Block
			1156	1	Screw Thrust Washer	28105-A	2	Universal Joint Pin
1049-A	1	Universal Joint Block	1157	1	Housing Thrust Washer	28110-A	1	Universal Joint Bolt, Complete
1050-G	2	Universal Joint Block Bolt, Complete	*1159(v)	1	Main Drive Gear Housing Cover	28567	2	Vane Shaft
1056	2	Outside Rear Bowl Cover	*1160(v)	1	Main Drive Gear	28568	2	Jet Plate Set Screw
1062	12	Universal Joint Clip	*1161(v)	1	Main Drive Pinion and Shaft, Complete	28569	2	Vane Control Handle
1064	3	Universal Joint Block	1162	1	Main Drive Shaft	28586	2	Vane Control Screw
*1065(v)	1	Intermediate Drive Shaft—Hollow Section	*1174(v)	1	Front Tender Trough and Support	*29133(v)	1	Distributing Table
*1066(v)	1	Intermediate Drive Shaft—Solid Section	*1175(v)	1	Rear Tender Trough and Gear Housing	29153-A	1	Right Vane
						29154-A	1	Left Vane
1072-C	1	Crusher	*1176(v)	1	Trough Sheet	*29185(v)	1	Elevator Screw
*1076-A(v)	1	Rear Drive Shaft—Style No. 2	*1177(v)	1	Trough Wearing Plate	29186-A	1	Double Universal Joint Link
1078	2	Rear Drive Shaft Bearing Bushing	*1178(v)	1	Rear Drive Shaft Bearing—Front	*29193-A(v)	1	Back Conveyor Trough
1085	2	Main Drive Gear Housing Bushing	*1178(v)	1	Rear Drive Shaft Bearing—Rear	29200	2	Vane Control Screw Locking Nut
1087	2	Main Drive Gear Housing Cover Bushing	1190	2	Tender Trough Adjusting Plate—Front	*29206(v)	1	Elevator Pipe
1090	1	Main Drive Shaft Key	1190-A	2	Tender Trough Adjusting Plate—Rear	*29209(v)	1	Jet Plate
1091	2	Main Drive Shaft Thrust Button	1191	8	Tender Trough Adjusting Plate Washer	*29213(v)	1	Intermediate Conveyor Screw
						*29215(v)	1	Front Conveyor Trough
*1093(v)	1	Oil Box, Complete	1199	1	Gear Housing and Cover Gasket	*29254(v)	1	Air Sealing Apron
1096	2	Rear Drive Shaft Collar	1221	1	Inside Rear Bowl	*29255(v)	1	Elevator Pipe Ball Cover
1100	12	1" x 2¾" Hex Hd. Bolt and Nut	1222	1	Outside Rear Bowl—Right Half	*29259(v)	1	Single Pointer Jet Gauge
1101	6	¾" x 4⅛" Hex Hd. Bolt and Nut	1223	1	Outside Rear Bowl—Left Half	*29260(v)	1	Double Pointer Jet Gauge
						*29264(v)	1	Elevator Pipe Filler Block—Right
			1660	1	Cover Slide	*29265(v)	1	Elevator Pipe Filler Block—Left
						29292	1	Cover Slide

* Part numbers followed by (v) are subject to variation. When ordering repair parts use HT Stoker Catalog No. 58 and Variable Sheet.

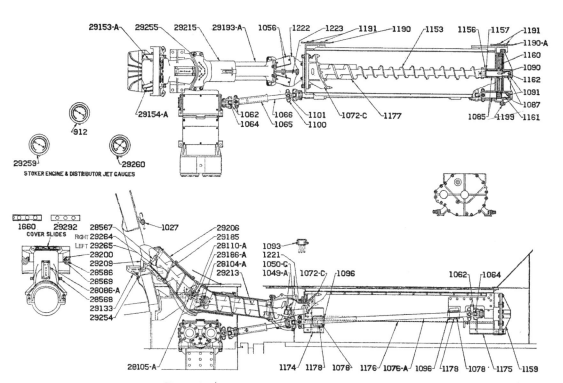

Fig. 11. Standard HT Stoker—Stoker Engine on Locomotive.

Pacific Type Passenger Locomotive, Class G-2-sa

Santa Fe Type Freight Locomotive, Class K-1-se

Northern Type Freight Locomotive, Class T-1

THE GRATES

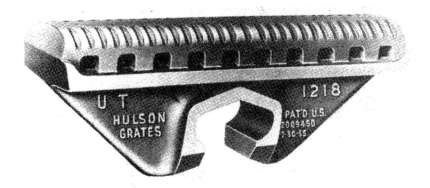

The Tuyere Unit for the 12" center Hulson Grate measures 2¼" wide by 11¾" long and weighs 8¾ pounds. It is available in any net air inlet from 12% to 26% of the grate surface. Made from close grained grey iron of 30,000 to 45,000 pounds p. s. i. tensile strength. The sections through the unit are uniform for even heat transfer. In service the units are cool. Heating occurs at stand-by periods, expansion is even and no distortion is developed as the unit is free to expand. Oxidation of the top surface is retarded indefinitely by the quick dissipation of radiant heat from the fire bed.

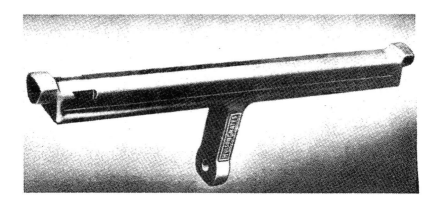

The carrier bar is a stubby I beam. Trunnions are 2¼" in diameter with bearing surface 1¼" long. A ⅜" steel pin is cast in the trunnion extending 2" into the base of the bar eliminating breakage or failure of the trunnion.

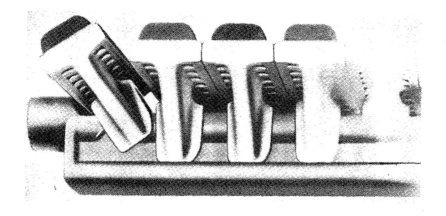

Units are applied at one end of the bar only. The last finger is "rolled" into place. Two-thirds of its bottom face is engaged by the carrier rib. Due to the generous bearing the unit cannot be dislocated to be lost under any operating condition.

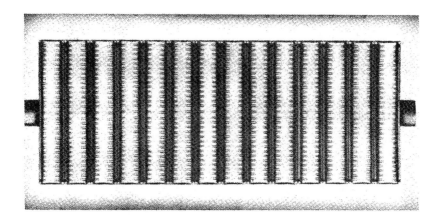

The assembled units present little "straight draft" air. The width (2¼") is controlled by the dry sand cores required for the tuyere openings eliminating any "sizing" or gauging when coming from the mould.

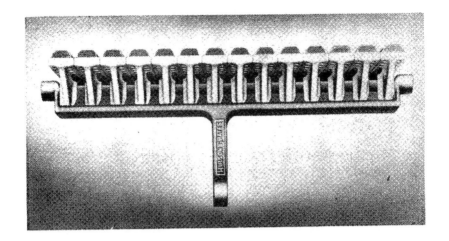

End view carrier bar and units. The valleys are wedge shaped and are cleared of any accumulation with the full throw of the grate.

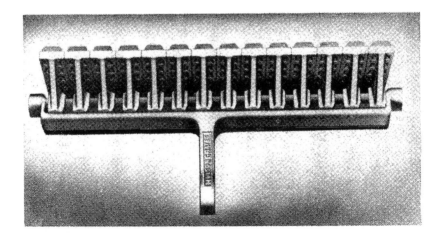

Bottom view carrier bar and units. Tuyere openings are generously tapered to the bottom. Any ash entering the tuyere falls through without plugging.

The three section grate assembly. All center and side frame castings are of grey iron. Failures of standard Hulson Grate center frame design are nil. Each section is provided with open journals the full length of the firebox on one side and closed journals on the opposite side. The closed journal, together with the connecting bar to the carries bar arms eliminates any lifting of the carrier bar from the journals under any operating condition. The weight of the 12" Unit assembly averages from 68 to 78 pounds per square foot, depending on the ratio of the length to the width of the firebox and the number of sections required in width.

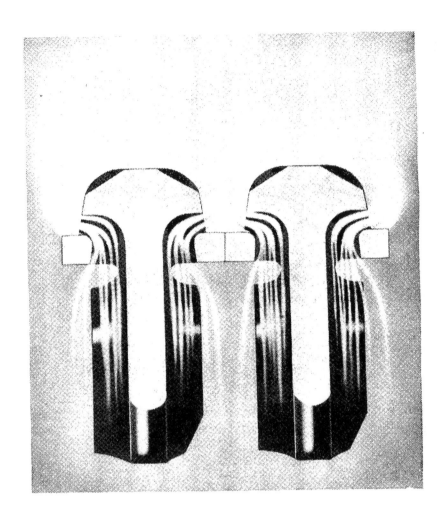

Cross section of two Tuyere Units illustrating baffled air admission. The design of the Unit Tuyere finger casting for the Hulson locomotive grate so arranges the tuyeres or orifices that each jet of incoming air is opposed by a similar jet resulting in a baffling action which reduces the velocity but not the volume of primary air. The result is a marked reduction in the lifting or tearing action on the fire from the sudden .increases in intensity of firebox draft when starting or at maximum combustion rates at high and continuous rates of working.

Light shaking or "wiggling" is necessary to control the depth of the fire. This illustration shows the shaker lever moved 10 inches from the normal locked position, which provides a free 2 inch opening between each grate of that section.

Shaking the grate becomes necessary when the fire is ne-
glected and ashes or banks are permitted to form on the
grate. This illustration shows the shaker lever moved 16
inches from the normal locked position which provides a
free 4 inch opening between each grate of that section.

Dumping the grate becomes necessary when clinkers are allowed to form on the grate or to dump the fire in case of an emergency. This illustration shows the shaker lever moved 28 inches from the normal locked position, which provides a total free opening of 5½ inches.

GRATES AND THEIR FUNCTIONS

INTRODUCTION

The proper use of grates requires careful consideration of two important factors. The first is to use the grates to keep the fire in good condition. The second is to use the grates so that they will not warp, burn, or break by misuse or carelessness.

The design of grate bars makes provision for the essential functions of the grates: (1) to get rid of the ashes, and (2) to permit air to pass through the fire.

The grates are usually made in six sections, and are shaken by shaker rods connected to a fulcrum lever. Each section is moved separately, and usually it is desirable to move the front grates first to make an even distribution of the draft through the fire.

The grate bars carry a number of sections so constructed that, when they are assembled on the bar, numerous openings permit the ash to pass through when the grates are moved. The size of the openings govern the amount of air passing through the grates in proportion to the grate area. The fire should be kept light and clean. A light, clean fire can be maintained by getting rid of the ash produced by the burned coal as fast as it accumulates.

Regardless of the type of grates, their chief function is to assist in keeping the fire light and clean. Keeping the fire bed light and clean prevents clinker formation and results in practically smokeless combustion.

To maintain a clean fire and to avoid smoke, the fire must be fed as little coal as possible and still maintain the correct boiler pressure.

The depth of fire and ash on the grates will gradually increase while the engine is being worked hard. This results in a dirty fire and clinker trouble. However, through proper use of the grates the fire can be kept as thin as the work being done by the locomotive and the grade of coal burned will permit.

Banks should be leveled before the stoker is started. If a bank forms after the stoker is started, the stoker should be shut off and the grates should be shaken under the banks to get through the thick area and then the stoker operation resumed. The use of fire hooks should be avoided, if possible. There should be no reason for using the fire hook if the fire is carried level and clean from the start and if the stoker speed and jet pressures are skillfully adjusted.

Stoker firing produces an ash clinker rather than a coal clinker because of the thin, hot fire bed that is maintained. Stirring the ashes up into the fire will certainly cause a clinker.

The stoker does the heavy work of firing a locomotive, but does not have the brains and judgment necessary to control the fire. The thin fire carried when firing with a stoker makes it necessary to use the grates carefully.

USE OF THE GRATES

There are three degrees of movement used by firemen when shaking grates: wiggling, shaking, dumping. All three terms, however, mean "shaking". Shaking the grates properly is a very important factor in keeping the fire in good condition. This requires GOOD JUDGMENT on the part of the fireman.

The three degrees of grate movements are:
1. Wiggling—light shaking—moving or rocking.
2. Shaking—half stroke or half open.
3. Dumping—full stroke or full open.

The FIRST DEGREE OF GRATE MOVEMENT which is the most frequently used method of operating the grates, consists of "moving" the grates often. This action scrapes the bottom of the fire bed and drops the heavy ash through the opening in the grate sections while the draft through the grates and fire bed carries the light ash out through the stack. The grates should be moved while the locomotive is working light. This combination of slight grate movement and light draft does not disturb the fire bed.

Keep the fire light and clean and increase the supply of coal if the fire gets too thin for the conditions under which the locomotive is working.

The SECOND DEGREE OF GRATE MOVEMENT consists of using little longer strokes on the shaker handle and "shaking" the grates to work large accumulations of ashes, small clinkers and slate through the grates.

This action reduces the depth of the fire rapidly and, because of the lifting action of the grates, has a disturbing effect on the fire. A short time should be allowed after shaking each of the grates to let the fire settle down before the next section is shaken to avoid disturbing too much of the fire at one time. Proper use of the first method of moving the grates will reduce the need for the second method to a minimum.

The THIRD DEGREE OF GRATE MOVEMENT is the dumping method which involves the use of the full length strokes and opens the grates several inches. This action is used when it is necessary to clean or dump the fire.

CARE OF THE GRATES

Ash accumulation in the ash pan should be kept away from the grates and shaker rod to avoid burning or warping them. A free flow of air through the grates keeps the temperature of the grates down, and therefore it is not necessary to maintain a bed of ashes on the grates to protect them from the heat of the fire bed.

The grate-shaker levers in the cab are provided with a locking device which assures that the grates are flat when the lock is in place. Keeping the grates flat prevents unburned coal from falling through the grates, and avoids burning the grate section. This might happen if an edge of the grate is turned so that it projects up into the fire.

HOW TO SHAKE THE GRATES

INTRODUCTION

The proper use of the grates is an important factor in maintaining a good fire and will eliminate considerable trouble if the correct procedure it followed. The movement of the grate will prevent the accumulation of ash, will maintain a clean fire, will prevent the formation of clinkers and thus will help control and equalize the draft. The grates are used to assist in controlling the depth of the fire.

The grates should be shaken only enough to keep the fire clean, to maintain the proper depth, and to maintain an even draft. Grates should be moved or rocked often instead of violent shaking at infrequent intervals. It is desirable to move the front grate first, if all sections are to be shaken, as changes in speed, the vibration of the locomotive, and the angle of the grates tends to cause the fire to work ahead and become thicker on the front grates. Grates should be shaken while the locomotive is being "worked light" because the shaking disturbs the fire.

PROCEDURE FOR OPERATING THE GRATES

1. Shut off the stoker to let the gases pass off before inspecting the fire.
2. Apply the blower to burn off the gases so that a complete inspection of conditions in the firebox can be made.
3. Examine the fire for evenness and thickness to determine whether or not the grates need shaking and which grates should be shaken.
4. Estimate the fire thickness by judging the distance between the top of the fire and the distributing tables.
5. Decide which grates to shake, how to shake them and when to shake them. The operation of the grates requires GOOD JUDGMENT.
6. Organize the fire and get the boiler pressure up if necessary before shaking the grate.
7. Shake the grate. Move one section of the grate, or more than one section as required by the fire condition.
8. Inspect the fire to see whether or not it is in good condition after shaking the grate.

OPERATING THE GRATE TO REMOVE NORMAL ASH ACCUMULATION

1. Place the shaker rod on the grate lever and move the front grate enough to correct any unsatisfactory conditions revealed by the inspection.

Caution: Be sure that the keeper locks are up and the shaker bar fits on the lever to avoid injury while operating the grates.

2. It is desirable to move the front grates first to make sure that air is coming through the front portion of the fire.
3. Control the fire depth and draft through the fire bed by gently moving the grate often enough to remove the normal ash accumulation.
4. Shake the rear grate.
  a. If it is necessary to shake the grate excessively to reduce the ash accumulation, it may be desirable to wait a short time after shaking each section before shaking the next section so that the entire fire bed will not be disturbed at one time.
  b. It may be desirable to hand fire light spots or holes if they develop while the grates are being shaken.
5. Restore the keepers to lock the grates in their flat position after the shaking operation is completed.
6. Shut off the blower.
7. Resume stoker operation.

Note: If the bad conditions were caused by an improper amount of coal or an uneven distribution of coal over the fire bed, make the necessary adjustment by changing the stoker speed or jet pressure.

8. Examine the fire to make certain that it is in good condition.

OPERATING THE GRATES WHEN THE FIRE IS THIN IN THE BACK OR HAS "WORKED AHEAD"

1. Notify the engineer if a bad fire condition is found so that he can "ease off" or, if necessary, close the throttle while the fire is being repaired. This is done to decrease the pull through the thin area which causes an UNBALANCED draft.
2. Equalize the draft through the grate area before attempting to build up the back of the fire, by placing the shaker rod on the front grate lever and by moving or rocking the grate gently under the thickest portion of the fire.

Caution: Be sure that the shaker bar fits on the lever and that the keeper locks are up.

Note: Shaking the grate violently disturbs the entire fire bed, fire is often shaken through the ash, and clinkers form on the grate.

3. Remove the shaker bar and restore the keeper locks to the locked position. This locks the grates in the flat position and avoids the burning of sections of the grate.

4. Shut off the blower.

NOTE: When the stoker is not operating, the fire dies down and the boiler pressure drops.

OPERATING THE GRATES TO ELIMINATE BANKS OR HIGH SPOTS

1. Determine the location and size of the bank and decide what movement of the grate is necessary to reduce the thickness of the bank and to restore the draft through that portion of the grate area covered by that bank.
2. Determine the cause of the bank while inspecting the fire.
3. Take immediate action to correct the distribution of the coal by adjusting the stoker jets.
4. Place the shaker rod on the grate lever that operates the section of the grate under the bank.
5. Move the shaker lever forward and backward until the high spot or bank has been eliminated and the fire has the proper thickness.

NOTE: Do not use the fire hook unless it is absolutely necessary. However, if the hook must be used, use it only to level off the bank, being careful not to tear the fire bed by digging into it with the hook.

OPERATING THE GRATES TO REMOVE CLINKERS

1. Determine the size and depth of the clinker while inspecting the fire.

NOTE: Frequent inspection of the fire will enable the fireman to detect clinkers before they affect the fire seriously. In general, the grate should be moved or "wiggled" often (not shaken), to insure free-flowing air through the grate and thus eliminating one of the chief causes of clinker formation.

2. Open the grate under the clinker far enough to break the clinker. This may let enough air through the area to burn the clinker out. If this cannot be done, apply one or more of the following suggestions.

   a. Break the clinker up and work it through the grate.
   b. Pull the clinker back under the distributing table or into the back corners with the hook.
   c. Take it out through the fire door, if possible.

CAUTION: Handle the clinker carefully to avoid getting burned. Use the hose if necessary and dispose of the clinker in a way that will not injure anyone along the right of way.

NOTE: If it is necessary to open more than one section of the grate to break up a clinker, a short period of time should elapse after moving each section, and care should be taken to save as much fire as possible.

THE LOCOMOTIVE BOILER

Reading
lines

92

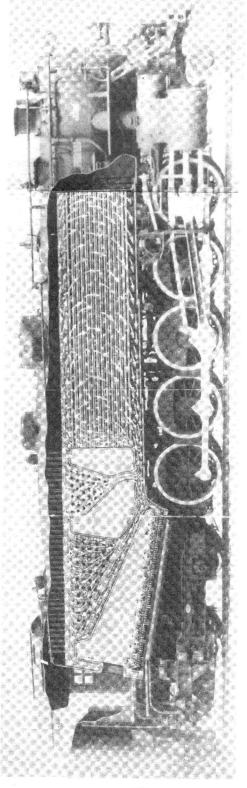

This cut-out of a modern locomotive shows the boiler equipped with syphon tubes. These tubes permit a free circulation of the boiler water, as shown by the arrows, improves the steam qualities and saves fuel.

THE LOCOMOTIVE BOILER

INTRODUCTION

There are several factors that must be considered with reference to the development of the ability of a locomotive to do work. However, fundamentally the engineer depends upon the steam pressure in the boiler.

Other factors are covered in different sections and the information in this chapter will be confined to a general description of the construction of the locomotive boiler with emphasis on the three main sections. Special consideration is given to the safety factors.

The responsibility of maintaining the proper water level in the boiler and the operation of the feed water devices should rest with the engineer. It is recognized that the engineer is responsible for the locomotive and its performance and it is his privilege to delegate some of his responsibilities to the fireman.

The fireman has three important responsibilities with respect to firing the locomotive boiler:

1. To maintain proper steam pressure by maintaining a good fire.
2. To be sure the water in the boiler is at the proper level at all times.
3. To prevent damage to the boiler by avoiding any condition which would cause rapid expansion or contraction of the boiler.

A sufficient amount of water must be evaporated in the boiler per hour in order to produce the required quantity of steam necessary for locomotive operation.

The locomotive fireman does not need to know all of the engineering technicalities and details of construction that go into building a locomotive boiler, but a knowledge of some features of the firebox, and smokebox is necessary if he is to get the full efficiency from the locomotive with a minimum of fuel. Poor judgment can cause damage to the boiler and firebox. The fireman must realize the importance of avoiding rapid changes of temperature which may cause severe expansions and contractions. It is particularly important to avoid doing anything or allowing any conditions to exist that will cause sudden changes in temperature in the boiler.

It is generally known that water boils in an open container at a temperature of 212° Fahrenheit. However, in a closed container, such as a locomotive boiler, the steam cannot pass off and, therefore, builds up pressure in the boiler. When more heat is applied, the steam pressure is increased.

The temperature at which water will boil rises as the steam pressure in the boiler increases, and consequently, more heat must be applied to keep the water boiling. The water temperature must be raised to about 298° Fahrenheit before the boiling point is reached when the boiler pressure is fifty pounds. The temperature must be raised to 380° Fahrenheit to boil water under 180 pounds pressure. These facts make it clear why a good fire and good boiler conditions are important in order to get sufficient heat to evaporate water under the working pressure of the locomotive when it is operating under various tonnage conditions and steam is being used.

THE BOILER

The boiler consists of three main units:

1. The enlarged back section which includes the firebox and combustion chamber.
2. The cylindrical section which contains the flues, superheater unit and steam dome.
3. The front section, often called the smokebox or front end which contains the draft appliances.

THE FIREBOX

The firebox is built into the rear section of the boiler and, except for the grates, is entirely surrounded by water. The heating surfaces of the firebox are those parts that are directly exposed to the fire, or heat from the fire, and are surrounded by water. These are principally the crown sheet, side sheets, back flue sheets, door sheet and combustion sheet. These sheets of the firebox transmit intense heat to the water and therefore, the most active generation of steam takes place around these heating surfaces.

The crown sheet at the top of the firebox is subjected to intense heat. It is also the part of the firebox that will be uncovered first if the water in the boiler gets too low. It is, therefore, important that the proper water level is maintained at all times to avoid any possibility of a damaged crown sheet. A damaged crown sheet creates a very serious condition. The fireman must be extremely cautious when observing and operating feed water appliances and be able to detect quickly any unusual boiler conditions.

A fireman should know that the WATER GLASSES AND GAUGE COCKS are in good working condition, and he should check them frequently so that he will know how much water is in the boiler at all times. He should be on constant guard against low water in the boiler while he is on a locomotive and should always examine the firebox, before and after a run for indication of defects.

A BRICK ARCH is built into the firebox. The arch is necessary because a large quantity of gases are given off during the process of combustion; the arch helps to burn these gases more completely by forcing them back over the bed of the fire. This action thoroughly mixes the air and the gases before they enter the flues and this promotes good combustion, saves fuel and prevents black smoke.

The arch also promotes considerable protection against leaking flues by causing a more even distribution of the heat over the flue sheet which forms the front of the firebox. The sudden draft of cold air, caused by a light spot in the fire or by opening the fire door, does not pass directly through the flues only, but is more evenly distributed by the arch.

SYPHONS AND ARCH TUBES

Most Reading Company locomotives have a brick arch built into the firebox. In addition to the brick arch, some locomotives are equipped with a syphon system and others with arch tubes. These devices assist in circulating the water in the boiler, provide additional heating surfaces and support the brick arch.

FLUES

The flues, attached to the back flue sheet at the firebox and to the front flue sheet at the smokebox, extend the length of the circular shell mid-section of the boiler. They carry the gaseous product of combustion from the firebox to the smokebox, in the front section of the boiler. The gases are very hot and play an important part in heating the water and in superheating the steam in the superheater units while passing through the flues.

A superheater locomotive has a number of large boiler flues in which steam superheating units are installed. Saturated steam, obtained from the steam dome, is passed through the dry pipe, to the superheater header and flows through the superheater units which are located in the large boiler flues. Steam, passing through the superheater units, is raised in temperature, obtaining a high degree of superheat from the firebox gases as they pass through the flues and out the stack.

CAUSES OF LEAKING FLUES.

Any condition that will cause the temperature of one area to change abruptly and be different from that of the surrounding areas will create strains which may cause leaking flues. If some flues do not heat as much as others, or cool more rapidly than others, they will exert a pull on the flue sheet. This strain tends to loosen the flues in the flue sheets. Plugged flues, uneven distribution of drafts, and sudden changes in firebox temperature, all put a strain on the boiler flues and have a tendency to cause leaks.

Some conditions that may cause leaking flues are the following:

1. A hole in the fire.
2. The fire door being opened too long.
3. Plugged flues.
4. Rapid changes in firebox temperature, especially if occurring repeatedly.
5. Using the blower too hard while cleaning the fire with the fire door open.
6. Injecting water into the boiler too rapidly when the fire is low.
7. Using the Worthington feed-water pump while the locomotive is idle.
8. Dirty or sooted flues.
9. Scale on the flues.

FIRE DOORS

The back head of the boiler carries the fire door or doors. They are operated by air and are controlled by a foot pedal. There is an adjustment of the exhaust plug to prevent slamming.

Air-operated doors can also be operated by a hand lever.

CAUTION: If it is necessary to place an arm or hand inside the door, the door should be latched or the door should be securely blocked open.

STAYBOLTS

Staybolts are used to support the inner and outer sheets of steam boilers and thereby prevent the sheets from being forced apart by the steam pressure. They are subjected to considerable strain because the inner plate gets much hotter and expands more than the outer plate. It is important for the fireman to consider the expansion and contraction that takes place when the heat in the firebox is increased or decreased and to control his fire to avoid rapid changes and uneven changes in firebox temperature.

Staybolts sometimes break and when they do they usually break at the outer sheet. A telltale hole is drilled in rigid staybolts under 8″ in length, which are subjected to severe stresses so that a broken bolt can be discovered by leaking steam or water.

Flexible staybolts are designed so that one end can move. This type is used where the sheets to be supported are subjected to considerable differences in temperature and where rigid staybolts would break frequently.

Hollow staybolts are generally used in many locations but especially where other devices are attached to the outside of the boiler and cover the staybolts; therefore, a telltale hole indication of a broken staybolt could not be easily detected. A hollow stybolt will give an indication at both the outside and the inside of a boiler if it is broken.

Broken staybolts, noted by the engine crew, should be reported promptly.

THE STEAM DOME

Locomotive boilers are provided with a steam dome to provide a space from which to obtain dry steam of the lowest moisture content possible.

SAFETY VALVES

Safety valves are automatic safety devices located on the top of the boiler. When the maximum boiler pressure, as indicated by the badge plate, has been reached, they automatically open and allow the steam to escape. Locomotives are equipped with either two or three safety valves which are set to pop off at different pressures with a difference of two or three pounds.

It is good practice to check the steam pressure gage with the badge plate and notice whether or not the safety valve pops at the maximum pressure. Popping can and should be avoided by careful firing and by proper boiler pumping.

THE SMOKE BOX

The smoke box, usually called the front end, contains the appliances which control, to a large extent, the draft through the firebox.

The burned gases from the firebox pass through the boiler flues into the smokebox and out of the stack.

The exhaust steam, as it passes through the cylinder exhaust ports and exhaust tip and out through the smoke stack, creates a vacuum in the smoke box. This produces the draft of air through the grates which is so necessary for good combustion in the firebox. The amount of draft depends on the amount of steam being discharged through the smoke stack. The amount of exhaust steam depends upon the conditions under which the locomotive is being worked. Thus, the draft through the fire bed is automatically increased and decreased as the demand for more or less steam varies.

The blower is used to create a draft when the locomotive is not being worked and no exhaust steam is passing through the smoke stack from the cylinders.

Steam leaks in the smokebox, caused by leaking steam pipes or nozzle-base joints, will permit steam to fill space in the smokebox, will destroy the vacuum and thereby decrease the draft through the firebox, causing the fire to burn red.

Air leaks in the smokebox can result in two bad conditions: 1, the draft through the fire bed is reduced, and 2, the smokebox may be burned or warped by combustion occurring in the smokebox itself. Air leaks must be promptly reported.

LOCOMOTIVE WATER SYSTEMS

LOCOMOTIVE WATER SYSTEMS

INTRODUCTION

Locomotive boiler efficiency is generally measured by the amount of water that can be converted into steam per pound of coal burned. There are several factors that have a direct bearing on the efficiency of the locomotive boiler. In the first place, the locomotive must be fired properly and equally important, the water must be supplied to the boiler at the rate it is needed.

Locomotives are equipped with a feedwater system designed to supply the boiler with water at a desired rate. Good, even pumping of the boiler is a technique which every engineer should develop. A considerable amount of coal can be wasted and the efficiency of the boiler lowered if the engineer does not have a thorough knowledge of the operation of the feedwater system and fails to maintain the water in the boiler at the proper level.

While the locomotive is idling or drifting, the water should be supplied to the boiler through the injector. The feedwater heater is used while the locomotive is working.

The temperature of the water delivered to the boiler also has a definite bearing on the efficiency of the locomotive boiler, naturally, water of higher temperature will require less heat to convert it into steam than water of lower temperature.

Water Level Indicating Devices

BOILER WATER LEVEL GAUGES

From the standpoint of safety and efficient operation, the devices on the back head of the locomotive boiler for indicating the level of the water in the boiler are among the most important devices on the locomotive.

The water level indicating devices include:
1. Water Column.
2. Water Glass, on the water column.
3. Water Glass, on the boiler.
4. Gauge cocks.

These devices are for the purpose of giving an indication of the height of the water in the boiler under all conditions of service so that the fireman and engineer will know that there is sufficient water above the crown sheet for efficient operation, and, especially for saftey.

THE GAUGE COCKS and THE TWO WATER GLASSES are devices provided for checking the water level in the boiler. The two water glasses provide a double check by the fireman and by the engineer. The GOUGE COCKS are used to check the water glass reading, and should be used frequently.

The WATER COLUMN is attached to the back sheet and gauge cocks and one water glass is mounted on it. This arrangement gives a much more accurate indication of the water level in the boiler than can be had when the gauges are mounted directly on the back sheet because the water has a tendency to "pile up" on the back sheet. This tendency to "pile up" is caused by the rapid circulation of the water when the locomotive is working.

The correct water level is two gauges of water, or even with the middle gauge cock which is midway between the upper and lower gauge cocks. The correct water level is indicated by the following tests:

1. A full stream of water when the bottom gauge cock is open.
2. A flutter of steam and water when the middle gauge cock is open.
3. A discharge of steam when the upper gauge cock is open.

Water glass connections may gradually become clogged. This condition is indicated by the slow or sluggish up and down movement of the water in the glass or by the failure of the water glass level to check with the gauge cock.

FEEDWATER DEVICES

LIVE STEAM INJECTORS

There are two types of injectors in use, the lifting and non-lifting injectors. The lifting injector raises the water and forces it into the boiler. The non-lifting injector is located below the lowest water level of the tank and the water flows by gravity to the injector.

WORTHINGTON FEEDWATER PUMP

Locomotives are equipped with either two injectors or with one injector and a pump. The feedwater pump saves fuel because it uses exhaust steam instead of a live steam to heat the water before it is forced into the boiler. Feedwater pumps are used to maintain the correct water level in the boiler except when the locomotive is drifting, standing or starting out, at which time the injector is used. Otherwise, it would be necessary to pump cold water into the boiler causing a severe strain to be set up. It would likewise reduce the steaming quality of the boiler.

Feedwater heating is a process for reclaiming a part of a

large loss. Previously, the exhaust steam which leaves the locomotive cylinders was used to provide a draft on the fire; the heat represented by this exhaust steam was a loss, as as far as its heat value is concerned. Therefore, by developing a method of using this exhaust steam to heat the water before it is forced into the boiler, a substantial saving is realized. Furthermore, the steaming qualities of the boiler are improved, the power of the locomotive is increased and the fuel is saved.

WORTHINGTON FEEDWATER HEATER
PRINCIPLE OF OPERATION

The principle of operation of the Worthington Feedwater Heater can be compared to that of any simple water pumping system. Water is pumped from the tender to the feedwater heater by a low pressure low speed centrifugal pump and the amount that is pumped into the heater is conrolled by a ball float in the heater. The water is mixed with exhaust steam, in the heater, by spraying it into a body of exhaust steam and it then flows to the hot water reciprocating pump which pumps it into the locomotive boiler.

Feedwater temperatures range close to within 10° of the temperature corresponding to the locomotive exhaust steam pressure in the heater. With 15 lb. pressure in the feedwater heater, the water temperature delivered to the boiler will be approximately 238° F. to 240° F. Temperatures increase with increase in back pressure, so that the highest feedwater temperatures are obtained when locomotive is working hardest and requires the most water. Less heat is therefore required from the fire box, the boiler operates at a higher efficency, and coal saving is derived not only from the heat recovery of the feedwater heater but also from operation of the boiler at a better heat transfer efficiency. The entire operation of this equipment is controlled by a SINGLE THROTTLE valve in the locomotive cab and the capacity can be regulated from maximum down to the smallest amount required. Usual variations in boiler pressures have no effect on pump capacity.

The cold water pump unit consists of a Pyle-National Steam Turbine and a Worthington Centrifugal Pump combined in one casing. The average speed of this pump when running at full capacity is about 3600 RPM; the steam which runs this pump is controlled by the float in the feedwater heater. The pump is lubricated by oil located in a storage oil cellar between the bearings.

Exhaust steam enters the heater through a group of exhaust check valves, and mixes with water coming through the spray valve. Water and steam mix in the approximate

103

proportions of five to one (five parts water, one part steam). As the heater fills with water, the ball float is lifted, which in turn reduces or shuts off entirely the live steam which operates the cold water centrifugal pump.

The hot water pump is the only high pressure part of the system, and being a long stroke generously built pump, it handles exceedingly hot water without difficulty. The principle of pumping water with a reciprocating pump is similar to pumping air with a reciprocating air pump. Both types of pumps have suction (receiving) and discharge (delivery) valves, and both types are double acting, taking and delivering water or air from both sides of the piston.

The cab gauge is connected to the cold water pump discharge pipe, and registers the pressure developed by the cold pump, in delivering water to the heater. This pressure is a combination of three resistances. (1) The resistance of raising water from tender level to feedwater heater. (2) Resistance created by the friction in discharge pipe, line check valve, and spray valve. (3) Resistance of locomotive exhaust steam pressure in feedwater heater against which the water is delivered. A change of pressure indication on the cab gauge DOES NOT indicate a change in the capacity at which the pumps are delivering water, but may only indicate a change in the locomotive exhaust steam pressure in the heater. Regardless of a change in gauge pressure, unless the pump throttle position is changed, there will be no change in the water pumping rate. Since the gauge indicates the cold pump discharge pressure, gradual movements of the gauge hand indicate changes in cold water pump pressure caused by a change in cold water pump speed. Since the cold pump pressure increases when the back pressure of the locomotive increases, the gauge will register roughly some 4 lbs. to 12 lbs. above the back pressure in the heater, depending on the amount of water being pumped.

The hot water pump obtains its lubrication from either a hydrostatic lubricator in the locomotive cab, or from a mechanical lubricator on the side of the locomotive and more recently through the use of individual lubricator operated by the pressure in pump steam cylinder. Water and steam conditions, as well as the size equipment, indicate the quantity of oil required. The minimum required can be determined by consulation with the Worthington Service Engineer.

A valve conveniently located in the cab is used in cold weather to provide a small amount of live steam, and direct it to the exposed portion of the equipment, to prevent freezing. This pipe has two branches, one which goes to the cold pump suction strainer box and one which goes into the cold pump discharge pipe. Both have ⅛ inch chokes at point where they attach to strainer box and discharge pipe.

Corrosive oxygen and other undesirable non-condensible gases are separated from the feedwater, and escape from the feedwater heater through vent pipes; one pipe running close to the front of locomotive stack, and the other running to the track under locomotive ash pan.

A small vent cock is attached to the gauge line in locomotive cab, and can be opened for venting air or steam vapor from the cold pump suction when priming cold water pump with water.

OPERATING INSTRUCTIONS
GENERAL

It is recommended that the pumping system be started as soon as the locomotive throttle is opened and the water pumping rate then regulated until both locomotive and pumps are working at a steady rate. Water should be fed to the boiler at the same rate it is being evaporated from the boiler, and a pump throttle position which satisfies this condition can be found by a small amount of experimenting with pump operating valve position. The pump gauge does not indicate the quantity of water being pumped, but it does indicate when the pumps are working.

PREPARATION FOR STARTING PUMPS

All valves in the system should be in proper working position, such as tank valve, boiler check, steam turret valve, etc. Lubricator should be feeding oil at a rate which is generally found to be satisfactory. Open cold pump air vent cock in cab to prime pump with water. If it is cold weather and suction anti-frost valve is open, close this valve before starting pumps —BUT CRACK IT OPEN AGAIN as soon as pumps have been started and are pumping water.

In order to get pumps started promptly, open pump operating valve one-half turn, and as soon as cab gauge indicates pressure, regulate pump throttle opening to that position which feeds water to the boiler at the desired rate. When starting up equipment, some seconds will elapse between opening of pump throttle and the indication of pressure on the cold pump gauge. This is because the hot pump must start first, and take some water from the heater, lowering the heater water level, causing the float to open steam control valve, and start the cold pump.

During running test, previous to locomotive leaving terminal, the sound of cold pump, and the pressure on cab guage, will rise and fall slightly and periodically, indicating the float control in heater to be working normally.

OPERATION OF PUMPS WITH LOCOMOTIVE USING STEAM

The amount of water pumped is entirely under control of the engineman and by proper adjustment of the pump throttle, any capacity rate can be obtained, while locomotive is working steam in cylinders.

CLIMBING HILLS

Worthington Heaters that have been applied in recent years have been designed with ample capacity to more than supply the locomotive boiler requirements. They have sufficient over capacity so that not only can they replace water that is being evaporated from the boiler, but also they can raise the boiler water level at the same time if desired. Since there is a large quantity of exhaust steam available while climbing hills, it is customary to pump as much water as possible at this time, and when the top of hill is reached there will be sufficient water in boiler to show in water glass after engine has tipped the hill and started down the other side.

COLD WEATHER PRECAUTIONS

CARE OF PUMPING SYSTEM ON THE ROAD (COLD WEATHER)

During freezing weather keep the anti-freeze pipe valve cracked open at all times except when pumps are to be started. Previous to starting pumps, the close anti-freeze valve and when pumps have started and are pumping water, the anti-freeze valve should again be cracked open. This results in the anti-freeze pipe being protected and also in furnishing a small amount of heat to the cold pump at all times.

During long drifting periods of the locomotive, this valve should be opened at least one-half turn and the pump throttle should be cracked open to keep hot water pump creeping. In the event that something occurs to the system which prevents engineman from using it to pump water, procedure as outlined above should be followed, and in addition all drain cocks on hot water pump should be opened.

If after a very long drifting period it is found that the suction hose has been overheated and pump will not handle water, shut off pumps, close anti-freeze valve for a short period and open "Cold Water Pump Air Vent Cock" in gauge line in cab. This will vent all steam vapors and pump will prime with water, after which pumps can again be started.

CARE OF PUMPING SYSTEM ON A DEAD ENGINE

In the case of a locomotive which has its fire dumped, either on the road or at a terminal, and the engine is to remain outside for a period long enough to permit boiler to get cold, the pumping system should be thoroughly drained. All drain cocks should be opened and investigated to see that they are not plugged with dirt. If equipment is drained while boiler is hot, extra care must be taken to prevent leaky boiler checks from permitting water to leak from boiler into pump.

A group of possible troubles are listed herewith; the causes of these troubles, tests that are used to uncover the cause of troubles, and what remedies are necessary to correct the troubles are also described.

PUMP WILL NOT WORK

TEST: Test pumps by opening pump throttle and carefully observing performance:

If cold water pump does not run:

1. Examine for restricted steam supply which may be located in stuck float valve, disconnected float, dirty steam line strainers, or plugged turbine nozzle.
2. Examine pump stuffing box, which may be pulled up too tight and binding shaft. There should be a slight water leak from this box, and pump shaft should spin freely with the fingers.
3. Examine turbine brake shoe, which may be binding in housing.
4. Turbine exhaust pipe should be run to atmosphere or point of low pressure and have no restrictions.

If hot water pump does not run:

1. Check steam supply to pump.
2. Examine steam valve gear for lack of lubrication, worn rings, broken parts, or stray iron which is binding reversing valve.
3. Examine cylinders for an obstruction.
4. Boiler check valve must be open.
5. Disconnect steam cylinder exhuast and check sound of exhaust for steam ring blow.

PUMPS RUN BUT WILL NOT PUMP WATER

TEST: Test pumps by opening pump throttle and carefully observing performance. Take particular notice to see that cold pump speed varies somewhat, which indicates proper movement of float valve.

COLD WATER PUMP

1. If cold water pump runs at an extremely high speed, and gives off a high vibrating tone, pump is not getting water. Tank valve may be closed, tank strainer may be dirty, tender may be empty, tank hose lining may be

loose and plugging inlet to pump, or pump may be air bound or over-heated with steam from anti-freeze valve.

Remedy:

If pump is air bound, STOP PUMP, open drain cocks beneath pump and ahead of line check valves. Also open cold water pump vent cock. Drain cocks must be open long enough to let all vapor out of pump casing while pump is not running, to permit pump to completely and properly fill with water.

HOT WATER PUMP

1. If cold water pump operates satisfactorily and there is plenty of water in heater, and hot water pump runs but does not pump water, examine hot water pump cylinder packing. If this is cause of trouble, there will be an unusually heavy flow of steam from both vents during running test.

PUMP WILL NOT SUPPLY BOILER

TEST: Test pumps by opening pump throttle and carefully observing performance. The drifting control valve should be moved to wide open position and hot pump run at full speed.

HOW TO PUMP A BOILER

INTRODUCTION

From the standpoint of safety and boiler efficiency the importance of maintaining the correct amount of water in the boiler cannot be over-emphasized. Knowledge of the correct method of pumping water together with the ability to interpret gauge indications is essential.

Fireman should be thoroughly familiar with the story which the water gauges tell and he should know when to add water and how to add it. The correct amount of water should be maintained in the boiler at all times from the standpoint of safety and boiler efficiency.

Fireman should recognize the fact that even pumping of the boiler is essential to good firing. Even pumping means operating the pump steadily at a speed just fast enough to keep the water in the boiler at the proper level.

The locomotive boiler is equipped with a water column, three water gauge cocks, and two water glasses. One water glass mounted on the water column and one is mounted directly on the boiler back head. These devices give the engine crew ample protection against an incorrect water level in the boiler. They should be tested and used according to instructions.

Running with an insufficient amount of water is a dangerous practice; on the other hand, over-pumping will cause the locomotive to "work water", which means that water will be carried into the cylinders with the steam. This destroys the lubrication of the steam cylinder parts and also causes an extra draft through the firebed which may be a cause for tearing holes in the fire.

CARE OF WATER SYSTEMS

TESTING THE WATER GLASSES AND GAUGE COCKS BEFORE LEAVING THE TERMINAL

1. Check the steam pressure by reading the steam gauge. The steam gauge should show within fifty pounds of the maximum pressure indicated on the badge plate when the locomotive is taken over by the crew.

2. Check the level of the water in the boiler by observing the water glasses and trying the gauge cocks. These should indicate two gauges of water and should be tested as follows:

a. Open the lower gauge cock—a full stream of water should flow.

b. Open the middle gauge cock—a flutter of steam and water should discharged.

c. Open the upper gauge cock—steam should be discharged.

d. Observe both water glasses—they should show half of a glass. The gauge cock should be used freqeuntly to check the water levels indicated by the water glass.

NOTE: BLOW OUT THE GAUGE COCKS WHILE CHECKING THE WATER LEVEL. Every possible attention must be given to keep the water glass valves, the gauge cocks, and the water column open and in good working order.

The two water glasses must be blown out when the crew takes charge of the locomotive.

3. Blow out the water column and water glasses by the following procedure:

a. Blow out the water columns by opening the drain valve wide until the water and sediment has been blown out of the column. Then close the drain valve. It is important that this be done first to prevent the sediment from entering the water glass.

b. Blow out the water glass on the column by closing the water valve and opening the drain valve. See that there is a good flow of steam from the drain pipe.

c. Close the steam valve and see that both this valve and the water valve are tight.

d. Open the water valve wide and see that there is a good flow of water from the drain pipe; allow both the drain valve and the water valve to remain open.

e. Open the steam valve until it is wide open.

f. Close the drain valve slowly and note that both water glasses indicate the same level of water as do the gauge cocks.

g. Blow out second water glass in the same way. The glass is mounted on the boiler and there is no water column to drain.

h. Make certain that the drain valves are closed tightly.

i. Make certain that the steam valve and the water valve to both water glasses are wide open.

TESTING THE FEED WATER PUMP

1. Check the tank valve on the left side of the tender to see that it is wide open.

2. Open wide the main valve on the turret.

3. See that the pump boiler check valve is wide open.

4. Open the pump operating valve slowly, and observe the delivery pressure indicated on the gauge. See to it that you understand what the different gauges indicate.

5. Run the pump long enough to make certain that the water is actually entering the boiler. The water level should be raised on the water glasses.

TESTING THE INJECTOR BEFORE LEAVING THE TERMINAL

1. See that the injector overflow valve is open.

2. Open the tank valve on the right side of the tender.

3. Open wide the main valve on the turret.

4. Adjust the water regulator handle to feed the desired water to the boiler without spilling out the overflow.

5. Shut off the injector by closing the steam and water valves when correct operation is assured.

6. Check the levels in the water glasses. The proper level should be two gauges or approximately half a glass.

PUMPING THE BOILER WHILE LOCOMOTIVE IS UNDER WAY

The fire should be in good condition before the locomotive starts out on a run. The injector should be used for the first four or five miles or about seven minutes after starting out. This would bring the temperature of the brick arch and firebox up to normal and will give the feedwater heater time to become thoroughly heated. The feedwater pump should then be started, and the injector shut off. The water pump should be regulated to run at an even speed which will be just fast enough to keep the water level in the boiler at two gages (approximately one-half glass). The best superheat is obtained by this method of pumping the boiler. GOOD EVEN PUMPING OF THE BOILER IS ESSENTIAL TO GOOD FIRING.

PUMPING OF THE BOILER IS ESSENTIAL TO GOOD FIRING

1. Check to make certain that the following valves were left open after testing the pump and injector at the terminal:
   a. The tank valves on both the left and right side of the tender.
   b. The main valve on the turret.
2. Open the pump operating valve slowly.
3. Regulate the water supply
4. Even pumping at a speed just fast enough to maintain the proper water level under all operating conditions is better than fast, intermittent pumping. Supply cold water to the boiler too rapidly and it will cut down the boiler efficiency.
5. Maintain correct water level. Look at the water glass. frequently to check the level of your water in the boiler. There should be "two gauges", or approximately one-half glass of water.

Smoke Control
and Draft Control

Reading
lines

SMOKE CHART
RINGELMANN TYPE

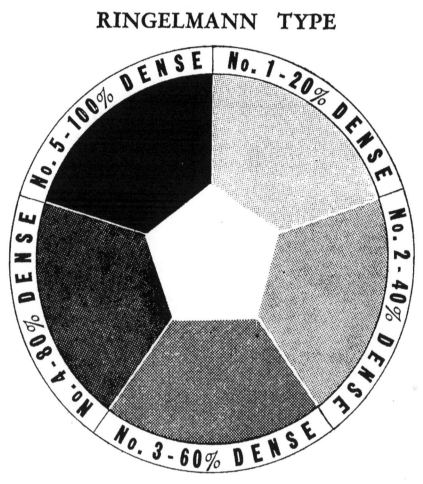

This miniature Ringelmann smoke scale will enable the observer to conveniently grade the density of smoke issuing from the stack.

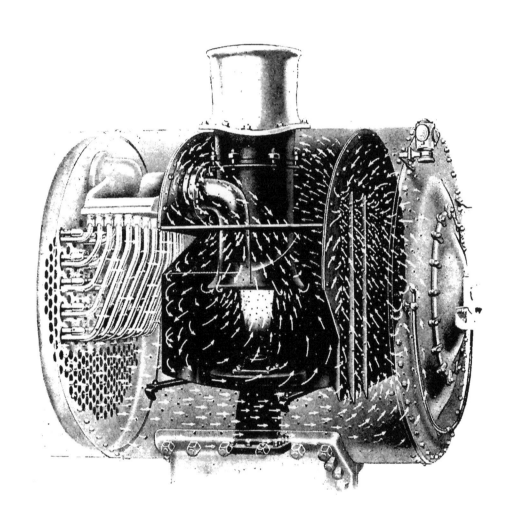

CYCLONE FRONT END
Arrows indicate movement of gases and cinders.

Smoke Control and Draft Control

INTRODUCTION

Excess smoke can be avoided if the fire is in good condition, if coal is supplied to the firebox in the proper amount, and if the coal is evenly distributed over the fire bed.

Black smoke is an indication of a waste of coal and is the result of overcrowding the fire, or in other words, feeding coal to the firebox in such a quantity that a sufficient supply of air cannot be obtained to burn the fuel.

The proper use of the blower when the engine is idle, and at other times when the throttle is closed, will assist in reducing smoke. The blower should never be used stronger than is absolutely necessary and should be shut off as soon as it accomplishes its purpose because it uses live steam directly from the boiler and burns out the fire. GOOD JUDGMENT in the use of the blower is essential.

DESCRIPTION OF THE BLOWER

The blower is a device used to create a draft through the fire bed when the locomotive is not working. It consists of three units: (1) A blower-control valve located in the cab. (2) A nozzle fitting in the smokebox. (3) A pipe which connects the blower valve to the blower fitting.

When the blower is turned on, live steam is discharged out through the smoke stack, thus creating a draft through the fire bed in the same way that the exhaust steam does when the locomotive is working and discharging exhaust steam through the exsaust nozzle.

USES OF THE BLOWER

The blower has three functions:

(1) To create a draft when building up or maintaining the fire before starting the run, and when long stops are made at the stations or on sidings.

(2) To keep gases and smoke from coming out into the cab when adding coal and when working on the fire.

(3) To control smoke. If the fire starts to die down when the engine is idle it may smoke unless sufficient draft is created with the blower to keep it burning. The blower may be used on a passenger train to lift smoke off the train if the locomotive smokes when the throttle is eased off.

WHEN TO USE THE BLOWER

(1) Use the blower when starting the fire.

(2) Turn on the blower before opening the fire door when the throttle is closed to avoid back drafts causing gas and smoke in the cab.

(3) Use the blower, if necessary, to assist in avoiding black smoke around stations, terminals, or yards, while maintaining the fire.

(4) Use the blower if the fire gets low after a long wait. It is better to keep the fire hot than to let it burn down too much and then have to build it up.

PREVENTING SMOKE WHEN PREPARING THE FIRE BEFORE STARTING

(1) Turn on the blower before starting to apply coal to the fire. The blower should be used lightly and then turned off as soon as the smoke is cleared up.

NOTE: When the locomotive is idle, the blower should be turned on before the fire door is opened so that the smoke and gases will not come out into the cab.

(2) Open the fire door and add coal by hand firing with the scoop.

(3) Check the boiler pressure.

NOTE: The fire should be in good condition and should have within fifty pounds of maximum steam pressure when the locomotive is turned over to the crew.

PREVENTING SMOKE WHEN THE LOCOMOTIVE IS UNDER WAY

(1) Inspect the fire frequently to be sure it is even and to make certain that the proper depth is being maintained. For information on how to inspect the fire refer to the unit "How To Inspect The Fire".

NOTE: Uneven distribution of coal restricts the draft at the high spots and results in incomplete combustion and excess smoking. Light spots intensify the draft causing a high carry-over of cinders and unburned coal which tend to cause plugged front-end netting.

(2) Watch the grade, size, and condition of the coal in the tender, and regulate the stoker speed and jet pressures accordingly.

(3) Adjust the stoker to run at an even speed and just fast enough to supply coal according to the steam demand.

NOTE: EXCESS SMOKE IS PRODUCED WHEN THE RATE OF FIRING EXCEEDS THE RATE OF BURNING.

(4) Adjust the stoker jets to the pressure required to get

116

an even distribution of coal over the firebox.

(5) Do not use the stoker to supply coal to the firebox when the locomotive is standing. Maintain the fire by hand firing to keep the boiler pressure up. Use the blower to assist in preventing smoke and to creats sufficient draft to maintain the fire.

Maintain steam pressure and have the fire in proper condition tö start out.

(6) Avoid over-crowding the fire at all times. A knowledge of the road ahead and the probable demand for more or less steam will enable the fireman to maintain a uniform boiler pressure by controlling his fire according to the demand for steam.

NOTE: Be especially careful to eliminate dense smoke in any smoke restricted territory.

ALWAYS OBSERVE THESE RULES

(1) Never use a blower STRONGER than it is absolutely necessary.

(2) SHUT OFF the blower as soon as it has completed its purpose. The blower uses live steam and is a heavy strain on the boiler.

(3) USE the blower WITH CARE if the fire is low or if there are holes in the fire. A draft of cool air may damage flues and staybolts.

THE SUPERHEATER

Reading Lines

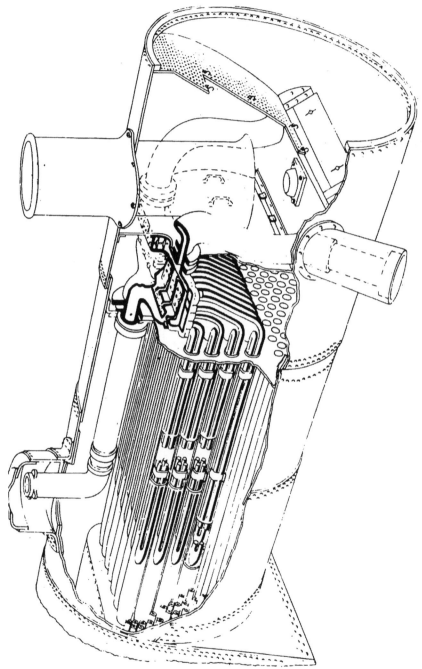

The superheater adds heat to the steam after it has left the boiler and is flowing to the steam pipes leading to valves and cylinders. A superheater does not increase the steam pressure, but it does increase the steam efficiency.

THE SUPERHEATER

The superheaters used on American Railroads all work on the same general principles. They add heat to the steam after it has left the boiler. A superheater does not increase the steam pressure in the boiler, but it does increase the steam efficiency. Superheated steam can be defined simply as steam having higher temperature than that corresponding to its pressure. The temperature difference between superheated steam and saturated steam at the same pressure is known as the degree of superheat.

Experience shows that it is only with a high degree of superheat—that the full benefit of superheating can be realized. Only with high superheat is there more efficient expansion, volumetric increase and minimum cooling of the steam by the cylinder walls during the working strokes, which together enables the steam to be utilized in the most satifactory manner.

The essential parts of a superheater are the header and the superheater units. The header is located in the smoke box and has two compartments. One compartment receives the saturated steam from the steam dome of the boiler and passes it to the superheater units. The other compartment of the header receives the superheated steam from the superheater and passes it to the cylinders. Most modern superheater locomotives, have the throttle installed in the superheated steam compartment of the superheater header instead of in the steam dome which was the usual location in older locomotives.

The superheater units are located within the boiler flues. Each unit consists of tubes joined together with machine forged turned bends so that each unit is complete with one connection into the saturated steam chamber, another connection into the superheated steam chamber of the header.

To be effective a superheater locomotive must have a hot fire in the firebox. A bright clean fire maintained at a minimum thickness and kept clean, level, and even in depth is essential. A hot fire is indicated when flames are bright and short. A low firebox temperature is indicated by long, smoky flames. Excessively bright spots and smoky flame indicate poor combustion and not enough draft through that area.

Printed in Great Britain
by Amazon